RUM

—THE MANUAL—

RUM
— THE MANUAL —
— DAVE BROOM —

MITCHELL BEAZLEY

This book is for Martin and Lisa,
friends always, and In Memoriam
to the genius of Dick Bradsell.

An Hachette UK Company
www.hachette.co.uk

First published in Great Britain in 2016
by Mitchell Beazley,
an imprint of Octopus Publishing Group Ltd,
Carmelite House,
50 Victoria Embankment,
London EC4Y 0DZ
www.octopusbooks.co.uk
www.octopusbooksusa.com

Distributed in the US by Hachette Book
Group, 1290 Avenue of the Americas,
4th and 5th Floors, New York, NY 10020

Distributed in Canada by Canadian Manda
Group, 664 Annette St., Toronto, Ontario,
Canada M6S 2C8

ISBN: 978-1-84533-962-3

A CIP record for this book is available from
the British Library.

Printed and bound in China.

10 9 8 7 6 5 4 3 2 1

Senior Editor Leanne Bryan
Copy Editor Jamie Ambrose
Proofreader Kathy Steer
Indexer Helen Snaith
Art Director Juliette Norsworthy
Designer Geoff Fennell
Picture Research Manager
  Giulia Hetherington
Picture Library Manager Jen Veall
Production Manager Caroline Alberti

CONTENTS

INTRODUCTION

When did it start? Maybe the Halloween game involving treacle scones, dipped in more treacle, suspended from the kitchen pulley. We'd try and eat them with our hands tied behind our backs. The swinging splatter of sticky blackness on face and lips was the first time I tasted molasses: sweet but bitter, the taste of iron and blood.

Rum would come later, in a fishermen's pub in Lochaline, drinking a mix we christened Rum, Sodomy, and the Lash. It wasn't exactly pleasant, but it made the motorboat ride home interesting. Then came venturing into Cadenhead's shop and buying something old from Guyana and being blown away.

Memories: drinking JB in beachside bars in Jamaica, avoiding cursed shopping bags in La Réunion, molasses adhering to my feet, the smell of muck pits, whirling machetes. I'll tell you the stories one day. You'll not believe most of them.

What they all involved was a combination of laughter, people, and a passion that showed how this drink was the pulsing heart of a culture. As my journey continued, so rum's complex tales of pleasure and exploitation became ever more fascinating. No spirit is as sticky with moral contradictions. Rum's story is as sweet and enlivening as a drop of syrup, as bitter as a treacle slap. It is a punchbowl filled with possibilities.

My rum life has been ... rum.

Entering this world is to sit at a domino match with opinions rather than flying tiles slamming onto the table, a place where things are debated long, loud, and hard. There will be laughter at the end. This remains a spirit with the ability to make everyone smile.

What have I learned? Rum had quality control when other spirits were in short pants. It was the benchmark for quality – not some inferior junk coming out of the Caribbean. That hasn't changed. It's time to appreciate this and redress the myths that have sprung up around rum. This is not just a world spirit; it is a world-class spirit and has been for longer than any of us realized.

It's time to celebrate rum's diversity and versatility. Charge your glasses. Let's fathom this bowl!

EXPLODING MYTHS

Rum is such a diverse category that it's no surprise some myths have built up around it. Let's try and disentangle some of them.

Myth 1: Sweetness

History shows that rum was sweetened from the start, so you can argue that this is traditional practice. My problem is that sugar levels in some rums are at quasi-liqueur-like levels. People may like it, but the downside of sugar doping is that any gain in popularity is a loss in complexity and character. Sugaring increases homogenization at the precise time when rum's diversity should be celebrated.

The practice is unfair to producers who don't, or are not allowed to, add sugar. Ideally, added sugar levels should be declared on the label and capped in the same way for cachaça or Cognac. There has to be honesty and transparency. However, given the problems over trying to legislate such a diverse category (see page 9), it might be easier for producers who don't add sugar or tint to say so on the label.

And another thing ... If you want vanilla notes in your rum, use first fill casks. If you want to flavour-up a rum, then call it spiced. Don't lie.

Rum is the world's most sociable spirit – never forget the fun.

Myth 2: Fun/No Fun

People smile when they think about rum. It's affability means it's an automatic choice as a party drink, but don't dismiss it as being only that. I was asked by a rum blogger if rum, in a bid to be viewed in the same light as single malt, should shed its fun image. "No!" I cried. Even now, in Scotch, new drinkers feel like they need a degree to start appreciating single malt. It's not a binary choice between being fun and being "serious". The former has to be retained while widening rum's remit. Rum covers more bases than any other spirit. Don't restrict it.

Myth 3: Age

Scotch has created a model where a spirit gains credibility at 12 years. This puts rum in an awkward position as it matures more rapidly in tropical climes. A lust for "Scotch-style" age statements in tropical-aged rums will tend to leave you with extract of oak in your glass. Rums don't get sweeter with age; they get drier.

Also remember, that while solera ageing (see p.56) is a valid technique, it needs to be explained that it's different to static ageing. It's impossible to give the exact age of a solera-aged rum as it's a blend of different years. You can only give an approximate, average age.

Solution? Don't look at the numbers; taste the spirit.

Myth 4: Labels

There is no universal, overarching rum regulation. Would it help if there were? Yes. Will it happen? Unlikely, as rum is already governed by regulations in each of its producing countries. What is legal in one place isn't allowed in another. Getting them to agree to another set of rules will be well nigh impossible. That means that reading a rum label is confusing at the very time that rum's premium sector is growing and people want to know what's in the bottle. It is detrimental to the spirit to confuse or obfuscate.

So, can we do a classification ourselves? Possibly. Luca Gargano of Velier has come up with a proposed classification system:

Pure Single Rum: 100 per cent pot (i.e. batch)-still rum.
Single Blended Rum: a blend of only pot-still and traditional column-still rum.
Rum: rum from a traditional column still.
Industrial Rum: Modern multi-column still rum.

If bars and retailers began to use something like this independently, then we'll at least have made a start.

Labelling, while often colourful, does need to be decoded.

HISTORY

It goes like this. Islands are colonized, planters arrive, they plant cane, import slaves, work out what to do with the molasses, and start shipping rum as a way to maximize the return from their crop. The deeper issue is one of identity: the symbiotic relationship between place and product that emerges in different ways at different times in rum's story.

Sugar cane is a type of grass.

SUGAR: THE SWEETEST SPELL

In the days before the time of humans there was sugar cane. One day, two bulges appeared in one cane's stalk. They grew and grew until the stalk split and the first man and woman fell onto the earth. According to Pacific islanders, we come from sugar. In chewing it, we taste our origins. It nourishes and supports; it is our generator. Our stories start with it.

Cane was domesticated in New Guinea as early as 8,000 BC, reaching the Asian mainland around 1,000 BC. In India, blocks of sugar syrup called *khanda* (where our word "candy" originates) were made around 500 BC. Sugar was heaven sent. Kāmadeva, the Hindu god of love, intoxicates his victims with arrows made of flowers, shot from a bow of sugar cane. The Buddha was said to be an ancestor of legendary king Ikshvaku (Iksu meaning sugar cane).

For Chinese Buddhists, sugar was both food and a medicine, while in its making – the removal of impurities – could be discerned as a metaphor for enlightenment. In the seventh century AD, Emperor Taizong of the Tang Dynasty sent expeditions to India to discover the techniques of sugar refining. Whoever tasted sugar fell under its sweet spell.

India – Home of the First Cane Spirit

Fermented "sugar wine" was made in both India and China, but there's no evidence that it was being distilled in China. In India, however, there are historical references to a sugar-based spirit. When Sultan Alauddin Khilji (r. 1296–1316) forbade the sale of wine, his subjects made a beverage out of readily available sugar, then distilled it.

Predating this, in Book II, Chapter XXV of the *Arthashástra*, an ancient Hindu treatise on politics and statecraft, is a list of the responsibilities of "The Superintendent of Liquor". Among the list of permitted beverages is *amlasidhu*, a spirit distilled from molasses. Although the original texts were written around 375–250 BC, it's likely that this reference comes from a later addition, but no later than 300 AD. Whatever the case, it would appear that cane spirit's origins lie in India.

HISTORY

Christopher Columbus' in-laws were in the sugar trade.

Persia

By this time, cane was spreading westward. It was first mentioned in 327 BC by Nearchus, one of Alexander the Great's generals, who wrote of "a reed in India which brings forth honey without the help of bees from which an intoxicating drink is made". This could be sugar wine.

Full-scale sugar-cane cultivation only began in the sixth century, when the plant was taken to Persia after the emperor Darius invaded India. While sugar was extravagantly used in cakes and pastries and as a medicine in Persia, there is no historical evidence that it was distilled.

When Islamic civilization spread across North Africa into Sicily and Spain in the seventh century, sugar cane came too. Egypt was a major centre of cultivation, while Spain would eventually have 75,000 acres under cane.

Cane and the Caribbean: the Early Days

The siren song of sugar is potent. It follows civilizations, helps found empires. This white gold was captivating, addictive, and expensive. Controlling production became increasingly important.

Portuguese Influence

In 1425, Portuguese Prince Henry the Navigator took sugar cane to Madeira and onward to the Cape Verde Islands, while the Spanish, also in expansionist mood, planted it in the Canaries. It was Canary Island sugar cane that Columbus (whose wife's family were sugar traders) took to the Caribbean on his second voyage in 1493. In 1552, Governor-general Tomé de Souza reported that slaves were drinking *cachaço* (today's cachaça): the first record of cane spirit being made in the New World.

Every plantation had its own still, and over the next 40 years sugar and spirit spread across Brazil. By 1640, it had reached the "Guianas" (now known as the Guyanas), where the Dutch had began to cultivate cane.

They weren't the first to see the area's potential. Sir Water Raleigh had sailed to "the large and bewtiful *[sic]* Empire of Guiana" in 1595 and 1617, reporting that "the soile besides is excellent and so full of rivers as it will carry sugar, ginger, and those other commodities which the West Indies hath".

The age of the sugar colonies had started.

A typical seventeenth-century sugar mill.

British Influence: the Enchanted Isles

On February 20, 1627, 80 colonists and 10 slaves stepped onto the empty beaches of Barbados. After the initial crops failed, they turned to cane, probably supplied by Dutch planters in Brazil, who also supplied stills and, possibly, distilling expertise.

Within two decades, Barbados became home to 75,000 planters, servants, and slaves, and by the end of the century it had become Britain's wealthiest colony. It was a place of fascination for visitors such as Richard Ligon, who wrote a detailed account of his time spent at William Hilliard's plantation between 1647 and 1650. Ligon was dazzled by the island's beauty, amazed at its fecundity, but knew that there was a worm eating at the rotten heart of this enterprise.

The British were to use cane to forge a new empire based on trade and exploitation. By the end of the seventeenth century, sugar planting was more advanced than any agricultural enterprise in Britain. For it to grow, the planters needed workers. Many were Irish and Scottish prisoners of war, or men press-ganged ("Barbadosed") into service. The island was a microcosm of British society, with a landed gentry building grand houses such as St Nicholas Abbey and Drax Hall, a working class, and, in time, an underclass of slaves.

The Caribbean soon became home to exploitation colonies.

Rum was being drunk but, tellingly, Ligon places it as seventh on his list of the 10 beverages most widely consumed in Barbados. It was still a booster, a medicine, a salve to the pains of toil.

The maniacal desire to make as much money as possible out of sugar resulted in Barbados quickly becoming a monoculture, one that was totally dependent on imports, even for fuel, from the American and Canadian colonies. The currency? Molasses and rum.

The English (as it was then) empire was based on commodities made in English colonies being sent to the mother country, where they could be sold on, or refined. All profits were centred there. The colonies, for their part, could only trade with England.

Sugar and rum became the empire's fuel. Planters began to move from the already tired soils of Barbados to St Kitts, Nevis, Montserrat, Antigua, and Guyana, and, from 1655, to the fertile island that would become sugar colony Number One: Jamaica. The plantocracy had arrived.

Rum was becoming increasingly important to the colonies' economies. As Jamaica's governor, Sir Dalby

THE BOOMERANG EFFECT

Glasgow had established links with Barbados soon after colonization, giving rise to a forgotten development that helped start the city's (and Scotland's) love of rum. In 1667, the Wester Sugar House started operating in Candleriggs, soon followed by a distillery. More were to come, including the Easter Sugar House. While imported rums would have been consumed, rum drinking also started with the consumption of locally distilled spirit. By the start of the eighteenth century, distilling, rather than sugar refining, was the main activity of the sugar houses. The same would happen in Bristol, London, and Liverpool.

Thomas, wrote in 1690, "We must consider the spirits arising from Melasses [sic]... which, if it were all turned into spirits, would amount annually to above £500,000 at half the price the like quantity of brandy would cost."

Rum Comes of Age

The eighteenth century saw Britain becoming a rich nation, thanks in part to the Caribbean. It also became a rum-drinking one. In 1697, a scant 100 litres (22 gallons) of rum were docked in England. By the final quarter of the century, it would account for 25 per cent of the spirits consumed.

Bristol was the first major rum port, trading initially with Barbados. Throughout the eighteenth century, almost 60 per cent of the city's trade was from the Caribbean, peaking in the 1780s, when a reluctance to dredge the narrow Avon Gorge meant that ships couldn't reach the port. By then it was playing second fiddle to London.

The Rise of Middle-class Imports

The eighteenth century was a time when "modern" society started to develop, and imported goods were the beneficiaries. In terms of spirits, this gave the new middle class three options: French brandy, Dutch genever, or West Indian rum.

The first suffered from high import taxes and outright bans. The image of genever was tainted by the cheap gin flooding the slums of London. Rum, however, was clear of all such negative associations. As Frederick Smith writes in *Caribbean Rum*: "The strategy of French wine and brandy makers ... was to regard rum as the drink of slaves [while] the British Caribbean interests attempted to market rum as an exotic drink of the nouveau riche."

Rum was everything gin wasn't: imported, aged, expensive, and made by the richest men in the land. All of the 40 major planter families had one family member in Parliament, resulting in a powerful lobby. The "Gin Craze" of 1720–60 played into their hands. In 1733, a pamphleteer wrote: "I believe all mankind will allow that scarcely a wholesomer Spirit can be distilled than that call'd Rum." By the middle of the century, aged Caribbean rum was more popular than cheaper alternatives, such as Medford rum from America.

Poor grain harvests in the 1850s also played into rum's hand. William Beckford, whose family were the most powerful planters in Jamaica, was also an MP and Lord Mayor of London. With his sugar lobby colleagues he successfully got a bill passed outlawing grain distillation. People turned to rum, and consumption

RUM AND THE NAVY

The British Navy was more significant than pirates to rum's development. When at sea, drinking water turned to slime and beer went sour. Drinking drams of neat rum soon became an accepted part of the sailor's daily routine in the West Indies. It also led to ill-discipline.

In 1739, Vice-Admiral Edward Vernon took command of the West Indies station. On August 21, he ordered that "the respective daily allowance of half a pint... be every day mixed with the proportion of a quart of water to every half pint of rum... in two servings a day". He also

recommended that the ration be augmented with fresh lime juice (to help combat scurvy) and sugar, "to make it more palatable to [the crews]". Should the Daiquiri be called the Vernon?

Much of the industry's early growth came as a result of the Navy's increasing requirements. Purchasing was centralized at the Admiralty in London via a preferred supplier, ED & F Man, which bought either direct or through brokers. The rums were transferred to linked blending vats in the Royal Victoria Yard, Deptford, which operated as a sort of solera system (*see* p56).

What started as a Jamaican or Barbados-based rum was, by the nineteenth century, predominantly a Demerara-based rum, with some lighter rums from Trinidad and Barbados. This in turn gave rise to the "dark" or "Navy" brand that was made by British blenders.

By the 1970s, it was felt that a daily draught of rum didn't exactly chime with the running of a modern navy and on Friday, July 31, 1970, 230 years of tradition ended.

"THE SLAVE GRINDING AT THE MILL"

The foundations for the barons' great houses, and Britain's prosperity, were built on the backs of slaves. Nearly 1.5 million African slaves were imported into the British West Indies from 1627 to 1775, and four-fifths remained there. Jamaica alone took half of the slaves in the first half of the eighteenth century.

They were bought through barter – not just with rum, but with other goods from England: linen, cutlery, guns, and cotton. The rum cities of Liverpool, Bristol, London, and Glasgow were also slave ports whose prosperity was directly linked to the trade, as were manufacturing cities such as Manchester. Much of the American rum industry was dedicated to supplying high-strength "Guinea rum" for bartering purposes. Slaves generated the profits. They also made the rum.

rose, especially in Ireland which, between 1766 and 1774, drank more rum than England and Wales.

Although the bulk of rum in Britain was from Jamaica (Barbados still traded primarily with the American colonies), from the 1740s a wider selection from the other sugar colonies was beginning to trickle in. In 1744, British Guyana (aka Demerara) had seven plantations. By 1769, there were 56. The Seven Years War (1756–63) saw the British temporarily take Martinique, Guadeloupe, and Cuba and start commercial distilling in all of them.

The sugar colonies were economically vital for Britain. By the end of the century, Jamaica's sugar plantations made £15m ($21,250,773) a year – five times more than any other colony – and the average (white) Jamaican was worth 20 to 30 times more than his equivalent in Britain.

It wasn't to last. The next generation of sugar barons had a *laissez faire* approach to their holdings. Plantations were placed in the hands of factors and began to fall into decline.

The problem was also cultural. The sugar barons were British, not Caribbean get-rich-quick opportunists. As Matthew Parker writes in his fascinating account, *The Sugar Barons*, "The West Indies had none of the things that sustained and nourished the northern colonies: a stable and rising population, family, long lives, and even religion. Instead there was money, alcohol, sex, and death."

The exploitation of the colonies was becoming entirely unsustainable.

RUM IN THE NORTH AMERICAN COLONIES

Rum was America's first spirit. Almost as soon as imports from Barbados started, molasses was being distilled, first in 1640 in Staten Island, New York, then in Boston three years later. By 1750, there were 25 distilleries in Boston and Rhode Island, 20 in New York, and 17 in Philadelphia.

The incentive to distil was strong. Imported brandy was expensive and grain was needed for bread. Molasses could be bought for one shilling a gallon, while rum sold for six shillings a gallon.

HISTORY

Rum was drunk all day, every day in colonial times: at home, in taverns, and in various (mostly simple) combinations. Bombo was a weak, cool drink of rum, sugar, water, and nutmeg; make it stronger and you'd have a Mimbo. Calibogus was equal parts beer and rum; a Cherry Bounce was cherry-flavoured rum; Manathan was a one to four sweetened mix of rum and beer, while a Stonewall mixed equal parts of rum and (hard) cider.

The greatest of these early drinks was the Flip. A huge mug was filled two-thirds with ale; then sugar, molasses, or dried pumpkin was added; then a gill of rum. This was then stirred with a red-hot poker to make it bubble and foam.

An extravagant spin on the Flip was the Bellowstop, a speciality of a tavern owner in Canton, Massachusetts, which involved mixing a pint of cream, four eggs, and four pounds of sugar. Ale was added to the Flip mug, then four spoonfuls of the cream mixture, then the rum – and finally the poker.

Rum was used for barter – and to subjugate the First Nations.

Rum fed the fur trade, was used as currency, and helped to subjugate North America's First Nations while preventing them from allying with the French. It was also consumed in large quantities as a bracer or "antifogmatic", and as a social drink (*see* box, left). As Wayne Curtis comments in *And a Bottle of Rum: a History of the New World in Ten Cocktails*, if London had its Gin Craze, then America wasn't far off having a similar one for rum at the same time.

"No Taxation Without Representation!"

More manufacture meant an increased need for molasses. The colonies were meant to trade with the British Caribbean, but in 1713, France banned the importation of rum and molasses, resulting in a large cheap supply of the latter. American distillers, now led by those based in Medford, Massachusetts, leapt at this opportunity to make their business more profitable. To counter this move, in 1733, the British introduced the Molasses Act, placing a high tax on the raw ingredient – and consequently, finished rum.

Rum was now entwined within an increasingly febrile political landscape. The colonists – who, unlike the sugar barons, didn't have representation in British Parliament – felt they were being unfairly targeted and ignored the Act. Smuggling increased. In 1735, only £2 of duty was collected.

The Sugar Act of 1763 made things even worse. While duty fell, the sugar barons wanted their trade back, and Britain needed cash to pay for the Seven

PUNCH

No matter where you were in the eighteenth century, the most common rum drink was that most democratic, hospitable, and convivial of libations: punch. It was a drink for coffee houses, gentlemen's clubs, country houses, and taverns, a tipple for the literati, the grand, the politicians, and the people – if you could afford it.

By the end of the century, however, a change was afoot. In the 1760s, in his London punch house, James Ashley started selling individual servings of punch – a drink that would morph into the cocktail. For more on punch, *see* pp.188–91.

Years War. The colonists watched as the Navy and domestic authorities aggressively enforced an Act that would have bankrupted the American rum industry – which by then was making 18,169,977 litres (4.8 milion gallons) a year in 143 distilleries.

Rum was now a focus for dissent. Drink rum and you resisted the colonial master, drink rum in a tavern and you met like-minded folk. The "bowl of liberty" was being fathomed.

A rum-induced psychological shift had taken place. By the 1770s, Britain was "The Other", while America was home. Unlike in the West Indies, in North America there was an identity with place. Rum fuelled rhetoric. Its taxation was one of the sparks that lit a war of independence, and when Britain had to decide whether to keep its sugar colonies or America, it chose the former.

This was rum's high point in the New World. Shortly afterward, in 1790, US First Secretary to the Treasury Alexander Hamilton imposed a tax on molasses and British Caribbean imports (who said Americans had no sense of irony?). Rum was immediately seen as the drink of the old regime. A new country deserved a new drink – made with its own produce, by its own people.

It needed whiskey.

HISTORICAL PRODUCTION

A study of rum's detailed early production records shows a spirit that was far from being an afterthought of sugar-making, but one for which deliberate quality choices were being made from considerably earlier in its timeline than most other spirits. Rum was not only already at Spirits University, it was writing the curriculum.

Richard Ligon's diagram of a rum distillery in seventeenth-century Barbados shows a room with space for two stills and a cistern, which probably acted as a fermentation vessel. On the island of Martinique, things seem to have been even simpler, if du Tertre's drawing of a *vinaigrerie* with one small pot with spout and worm, is accurate. Rum was precocious, however, moving rapidly from being the rough drink for slaves, to punch for planters, then a desirable export product, whose sale could offset the running costs of the plantation.

After fermenting for 24 hours, molasses was added. This made up six per cent of the total volume and was added in two stages: three per cent after 24 hours, then the remainder a day after "or when the wash is rich and in a high fermentation". Martin advised adding all the molasses at the same time for a more controlled fermentation, though this older system continued to be used. Temperature control was used in the form of buckets of cold water if the (week-long) ferments rose "to near blood heat", or by adding hot water if it got "sluggish".

Eighteenth-century Rum: From Scum to Specialization

In 1707, botanist Hans Sloane reported in *A Voyage to the Islands...* that rum was being made from "Cane-juice not fit to make Sugar... or of the Skummings of the Coppers in Crop time, or of Molossus and water fermented about fourteen days in Cisterns". In other words, early rums were distillates of the scummy froth taken off the sugar pans. Molasses would in time become part of the fermenting ingredients, but skimmings would always also be there, with more being used at the beginning of the season. When sugar-making was over, more molasses would be used.

There was one other ingredient in this mixed wash that first appears in an account of rum distilling given in 1707 to the London distiller William Y-Worth by "a Person of Ingenuity": "...in Barbados... they take the Molasses, foul sugar and their canes and ferment them together with the remains of the former distillation....." This is the first mention of the use of dunder (*see* p.24).

The eighteenth century is full of tracts and instructions to planters that illustrate how quality was as important in their distilleries as it was in their sugar estates. This second generation of planters wanted to make money, and that meant a forensic understanding of what was happening and how it could be, in that great eighteenth-century term, "improved".

A key text was written in 1754 by Antiguan planter Samuel Martin, whose Greencastle estate became a quasi-university college of sugar cane. Martin's *An Essay Upon Plantership* is both a summary of best practice and canny advice. Cleanliness is key to his approach, along with cooled, filtered, dunder, ("lees") used "as yeast or barm" to start fermentation, temperature-controlled fermentation, and slow cool distillation. Much of this comes from his research into Barbadian planters, "the best distillers in all the sugar islands", who were producing a double-distilled "cooler spirit, more palatable and wholesome" compared to the higher-strength spirit made in Jamaica (possibly by triple distillation), which was "more profitable for the London market because the buyers there approve of a fiery spirit which will bear most adulteration".

By this time there was also an understanding of what we now call terroir. In his 1774 *The History of*

A rum plantation in Antigua as described by Antiguan planter Samuel Martin.

Jamaica, Richard Long reports that the rich soils of Jamaica's north coast produced a syrup "so viscid, that it often will not boil into sugar; but these estates produce an extraordinary quantity of rum. The south side lands, on the contrary, produce a less proportion of rum, to a larger quantity of sugar..." Specialization had begun.

By 1794, Bryan Edwards writes of a new "improved" method of rum-making in Jamaica, which upped the dunder levels to 50 per cent. Without dunder, distillers would have to add "most powerful saline and acid stimulators" and risk over-souring the ferment. What his and Martin's work show is an understanding of dunder that predates Dr. James Crow's use of sour mash in bourbon by 30 years.

The Nose D'Void of Funk

The battle with rum's odour started early. For Hans Sloane in 1707, it was "an unsavoury Empyreumatical scent" while Y-Worth talks of how rum "carries with it so strong an Hogo". The solution was redistillation, – or, Sloane says, "mixing Rosemary with it."

UK DISTILLATION METHODS

According to Ambrose Cooper, writing in 1757, British molasses spirit needed "fresh wine abounding in tartar" being added to the fermenter to increase the acidity. This was, at least, better than Y-Worth's solution of 50 years earlier, which kick-started the fermentation by adding "a pot of very strong mustard, with a Horseradish a good Onion and the value [white] of an Egg".

For Cooper, adding sweet spirit of nitre (aka ethyl nitrite) gave vinosity, and the subsequent spirit could "be made to pass on ordinary Judges for French Brandy" It was predominantly used for adulterating rum, arrack, and brandy. Sugar spirit was "extracted from the washings, scummings, dross, and waste of a sugar-baker's refining house", double-distilled, and also used mainly as an adulterant.

Dealing with the funk, the hogo, the "empyreumatic" smell, the "stinkabus" was rum-makers' greatest issue in the eighteenth and nineteenth centuries. For Y-Worth, the hogo was the result of "operators often using the remains of their distillations or wash... for beginnings instead of liquor" – i.e. it came from dunder. But as dunder made up most of the ferment in those days, the issue was how to control its effects.

Taming the funk led to the concept of "light rum" which, in Cooper's mind, would have been better "for common use of making punch... as the taste would be cleaner. In this state it would nearly resemble arrac". Now there's a telling line.

By the end of the eighteenth century, distilling consultants were arriving in the islands to improve rum further. One such was London scientist Bryan Higgins, who worked in the southwest of Jamaica for three years from 1797. His analysis centres on eliminating the "acetous ether", which came across in the early runnings of a still. His advice was either to take a middle cut from the first distillation, or stop using those "tainting and oily products of the putrescent filth," (i.e dunder and skimmings), because that's where the problem lay. Using molasses, he decided, would be the best option – pretty much what Y-Worth had concluded almost a century before.

Higgins's conclusions weren't taken on board – in Jamaica, at least. Only two years after his report, merchants in New York were still taking the bungs out of casks of Jamaican rum to try and clear them of the aroma.

The armies of Sir Nose D'Void of Funk were on the march. It would be the successful creation of "light" rum in the nineteenth century that would change the face and the aroma of rum forever.

The Nineteenth Century: Searching for the Grail

In the nineteenth century, technology, a change in the consumer palate, and a desire to retain, or establish, terroir split rum into multiple camps.

The first changes were, inevitably, linked to sugar. As the industry suffered, planters had to decide whether to stick with sugar or go with rum. While Jamaica stuck with dunder-rich styles, they were beginning to disappear elsewhere in the Caribbean

HISTORY

MATURATION

Sloane talks of rum being allowed to "stand under Ground in Jars" as a way of removing any taint. For Cooper, rum "must be suffered to lie for a long time to mellow before it can be used". In other words, maturation was being used for quality purposes. Again, this predates whisky.

It was obvious to anyone that the months in cask onboard ship changed, mellowed, and improved rum. In time, this mellowing became common practice, and "old" rum was more popular in the British market. Ten-year-old rum was on sale in London at the end of the nineteenth century.

For Higgins, maturation was, "highly preferable, even when recent from the still, to any new rum of the ordinary process". In other words, pot-still rum was best when it had aged.

Edouard Adam's still of 1801 was one of the earliest attempts at continuous distillation.

as consultants' advice began to be heeded. Now that distilleries were starting to operate independently from sugar mills, skimmings were no longer available. These differences were then amplified by the introduction of different types of stills.

Up until this point, everyone used pot stills. In Europe, however, there was a movement to make distilling more efficient. Batch-distilling was time consuming. A continuous system was the holy grail, where as long as a fermented liquid went in one end of the still, spirit would flow from the other. Engineers across Europe came up with a multiplicity of solutions, and most of their designs were shipped to the Caribbean.

Edouard Adam's still of 1801 placed two or three mini-stills between the pot and the condenser: the start of the pot and retort system. This would be refined throughout the nineteenth century and remains the most common method of making pot-still rums today.

Corty's Patent Simplified Distilling Apparatus (1818) was adopted across the Caribbean, with most in Tobago and British Guyana (which also imported Coffey and pot and retort systems). Water-cooled plates in the neck of the still increased reflux and increased the strength, yielding a rum that "does not possess its peculiar aromatic flavour in an equal degree with spirit of 30–35 per cent overproof". In other words, the funk had gone.

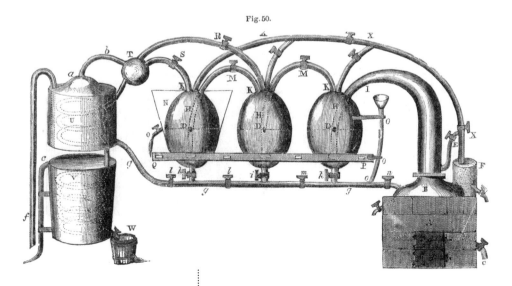

Fig. 50.

Cellier-Blumenthal's design was the forerunner of a number of column stills.

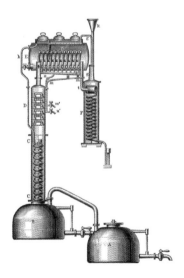

Cellier-Blumenthal cracked the problem of continuous distillation with the introduction of a pot (patented 1813) that led into a column divided into plates. This was adapted in 1818 by Parisian apothecary Louis-Charles Derosne, then by Dutch sugar-trader Armand Savalle, whose design would become widely used in rum production, as would Aeneas Coffey's linked column configuration.

"The different kinds of stills now in use are so very numerous, that it is quite impossible for me to name them," wrote Jamaican planter Leonard Wray in 1848. Politely rejecting Coffey stills for rum production he concluded: "I have never known any to surpass the common still and double retorts."

Significantly, all of these new lighter, rums were considered to be of a higher quality – and fetched a higher price on the European markets. Demerara's stock was rising, thanks to the adoption of new stills. There, according to Charles Tovey, distillation had been carried to "a high state of perfection... to be as much prized in the American market as Jamaica is preferred in the English market".

Jamaican pot still, the dominant style for over a century, was under pressure. Its neighbours were becoming bolder, defining their rums as not being Jamaican. To establish their identity they had to be different. For Jamaica, the situation was existential: what are we? How do we define ourselves?

Dunder Heavy Manners: the Fight for Jamaican Identity
Leonard Wray studied sugar planting in Asia, Natal, and the Caribbean (and introduced sorghum to the US). His book *The Practical Sugar Planter*, written in 1848, is an invaluable source for how quality Jamaican rum was made in this period. He urges young distillers to let their fermentations run long and slow: up to 10 or 14 days. His recipe contained skimmings, dunder, molasses, and water – and the contents of the muck pit, "an exceedingly noxious combination, from which the most loathsome and unwholesome emanations constantly arise". Yes, he had the funk, but he also filtered his rum post-distillation – and then sweetened and coloured it.

More insights can be gleaned from the diatribes written in the 1840s to the "Jamaica Standard" by another planter, W.F. Whitehouse, which not only give a detailed explanation of rum production, but also a sense of the politics of the time. I'm indebted to Stephen Shellenberger's www.bostonapothecary.com for bringing Whitehouse to my attention.

Like Wray, Whitehouse was a defender of tradition, but also keenly aware that quality improvements were needed. Much of his writings were rants against a consultant distiller called O'Keefe, who had tried to impose a new system of "scientific" production. Whitehouse devised his own method, challenged O'Keefe to a rum-distilling duel – and won.

Despite the Jamaican rearguard, by the middle of the century, quality in rum was seen as being higher proof, and funk-free. By the end of the century, light "Common Clean" rums were being made in Jamaica.

The Age of Science

The nineteenth century saw changes in people's palates, the arrival of cocktails, commercialization, and a new era for rum, with the building of a sense of identity.

As America switched allegiance to its own whiskey, the rum trade focused increasingly on Britain. In 1806, the 115-hectare (285-acre) Rum Quay opened in London's East India Docks. Rum merchants were now establishing themselves across Britain. Using their names as guarantees of quality, they blended marks from one or more countries. London had Lemon Hart's eponymous brand; White, Keeling had Red Heart; and Alfred Lamb has his Navy Rum. Liverpool had Sandbach Parker & Co, Hall & Bramley and many more, while Dundee merchant George Morton's Old Vatted Demerara (O.V.D.) and Old Vatted Jamaican Rum brands appeared in the 1830s and 1840s. Blending gave volume; it also provided consistency and complexity. The majority of these rum blends predate blended Scotch by 30 years.

Jamaican rum still dominated the British trade, but sugar planters were having to deal with major changes, the most significant being the banning of trafficking in slaves in 1808, followed by full abolition in 1833. Amazingly, the sugar lobby managed to convince the British government to pay them £20 million (US$27.7m) "compensation" for the loss of their

FILTRATION

Charcoal filtration, employed to remove impurities, had been used in vodka since the late 1780s, after Johann Lowitz's discoveries in Russia. By 1794, it was being used in Caribbean sugar production and then in rum. Wray outlines filtering through charcoal in his book. The biggest impact came in Cuba. In 1805, still designer Charles Derosne also invented a charcoal filtration system for the sugar industry. Both of his inventions were adopted by the burgeoning sugar industry in Cuba. The first plantation to install one of the Derosne defecators was Wenceslao de Urrutia's La Mella. Filtration of the spirit was the last step in the making of light rum.

GLOBAL RUM: INDIAN OCEAN

Although India can claim to have made the first cane spirit, its first dedicated rum distillery only opened in 1793, when William Fitzmaurice began to distil for the East India Company. In 1801, Carew & Co. built a dedicated rum distillery in Kanpur.

By the nineteenth century, Mauritius – which had already been distilling for almost 100 years – was as important in terms of volume as Martinique, supplying much of the Indian Ocean market and further afield. George Morton had Mauritian rums in his inventory at the end of the nineteenth century. Rums were by now also being made in Réunion, the Seychelles, Madagascar, and South Africa.

London's West India Docks was once the world's largest rum repository.

slaves. In order to get the bill passed, abolitionists had to agree that slaves were property after all.

With sugar now becoming more expensive to produce, and cheaper beet sugar appearing, the older sugar islands began to look increasingly to rum. By the end of the nineteenth century, rum was more important than sugar for Jamaican exports, and the island's producers were increasingly focused on what made their rums different: what, in effect, made them *Jamaican*.

The only sugar colony to grow in importance during this period was British Guyana, thanks to the mass importation of indentured workers from China and India. Each of its 300 plantations had its own distillery, and by 1849 almost half of the rum imported into Britain was Demerara. People's palates were changing.

It was a Liverpool firm that changed the face of British rum imports. Josias and George Booker settled in Demerara in 1815 and began shipping sugar and rum to their home city in 1835. By 1866, Booker Brothers & Co. was called "the principal shopkeepers of the colony". It merged with former employee John McConnell's firm at the end of the century. Rums were shipped on the firm's own "Liverpool Line" and sold to the brokers and merchants.

By the late nineteenth century, new technology was being used in rum-making.

As a result, Liverpool was fast becoming the most important rum city. It had expanded its docks throughout the eighteenth century, and kept its berthing fees lower than those of London, which encouraged merchants to move there.

British Guyana's volume and varied marks made it a blender's dream, and with increased investment in technology, Demerara rums grew in importance. By the end of the century, the Navy blend contained no Jamaican rum at all, and was predominantly Demerara based.

Rum's position was less secure in the US. Having fallen from favour, it missed out on the first flush of cocktails, while the Temperance movement used it as a catch-all term for debauchery, while the abolitionist movement inferred that rum drinkers were pro-slavery.

Its image was about to change, however, thanks to a new rum country with a new style: Cuba.

THE RISE OF CUBA

Cuba gave birth to what is now the world's most widely produced rum style, the one that was behind the great rum cocktails, the rum that saved the category and changed it completely by consigning the older styles – Demerara, Jamaican, Barbadian – to walk-on parts.

It took 300 years for the country to get there. For much of Cuba's history, distilling was banned. The Spanish Empire's wealth was to be built on gold, not trade. When no precious metals were found in Cuba, the Spanish foraged ever further into Mexico and Central and South America. Its holdings in the Caribbean became little more that staging posts.

Cane was scarcely planted in Cuba, so distillation barely happened. Then In 1714, a Royal Decree ordered all rum-making equipment to be confiscated and broken up in order to safeguard the production and sale of Spanish brandy and wine.

In 1762, the British captured Cuba and brought in 4,000 slaves, sugar-making and distillation equipment, and, just as importantly, experience. Although the occupation lasted only 11 months, it changed Cuba. When Spain regained control, it opened up trade and, in 1777, rum production was legalized. Spain's

GLOBAL RUM: THE PACIFIC

It wasn't until 1837 that the Philippines had its first independent distillery, San Miguel, owned by Domingo Roxas. In 1854, Ynchausti Y Cia followed (renamed Elizalde & Co. in 1893), by taking over the Tanduay Distillery. Today it makes the third bestselling rum brand in the world.

Although historically a large consumer of rum from its early colonial days, when rum served as currency, Australia had to wait until 1866 and the passing of the Distillation From Sugar Act before its own commercial industry began with Queensland's Oaklands Sugar Mill. Rum historian Chris Middleton has uncovered 18 rum distilleries operating in Queensland in 1890. The region also had the world's first (and probably only?) floating rum distillery, the *SS Walrus*, which distilled while floating around the coastal plantations near Brisbane.

Elsewhere in the Pacific, rums emerged from Hawaii, Fiji, Tahiti, and New Caledonia.

isolationist stance was failing. It was time to join the sugar and rum club, though it took two revolutions – the American War of Independence (1775–83) and the Haitian Revolution (1794–1804) – for the momentum behind a sugar industry (and therefore behind rum) to start building.

Cuba was ideally placed to supply newly independent America's sugar craving, particularly after the world turned its back on liberated Haiti. Mass importation of slaves occurred (abolition only took place in 1886) and by 1820, there were 652 sugar factories in Cuba. By 1829, the island was outproducing all of Britain's sugar colonies. In 1860 the island had 1,365 rum distilleries.

Enter Bacardí

A new wave of immigrants also arrived and modern distilleries were built in Havana, Matanzas, Cárdenas, and Santiago de Cuba. It was to the last city, in 1830, that a Catalan immigrant named Facundo Bacardí Massó moved and started to sell rum for an English distiller called John Nunes. In 1862, Bacardí bought Nunes' distillery and began to bottle his own rums.

The new Cuban distillers knew that they had to make a rum that was different. Slowly, distilleries became independent of the sugar factories, while improved technology allowed Cuban distillers to create a new, lighter, softer style of rum. Since its sugar industry was utilizing new technology, it was inevitable that Cuba became an early adopter of columns and pot-column hybrids. When filtration (*see* box, p.25) arrived, Cuban light rum was born and the island could differentiate its spirit from the rest of the Caribbean.

With this came a shift in attitude on the part of Cuban residents. In line with the plantocracy in the British Caribbean, the settlers in Cuba had seen themselves as displaced residents of the mother country. They had been raised on imported brandies and wines, while *aguardiente* was for slaves.

A refusal to drink what you make shows a certain detachment from place. Conversely, being proud of your own drink shows an attachment to that particular location. As the cause of independence began to gather pace, so the new Cubans begin to drink their new spirit. Rum became Cuban at a time when Cuba began to establish a true sense of self.

After independence in 1898, rum production increased, initially as a way to generate capital for the new country. Demand boomed. By now, Cuba was ahead of its Caribbean rivals in another aspect. While most of them were still producing bulk rum to be shipped and blended abroad, Cuban distillers were establishing that nineteenth-century phenomenon: The Brand. Whether it was Havana Club or Bacardí, producers owned what they made.

Rhum: the Antilles and *Agricole*

In 1635, the French settled Martinique, and in 1644, Benjamin da Costa, a Dutch Jew from Pernambuco, brought the first still to the island. The earliest accounts of distillation in the new French colonies come from Père Jean-Baptiste Labat, who lived in Martinique and Guadeloupe between 1694 and 1706. He was a man of contrasts: priest, slave owner, adventurer, anthropologist, planter, and distiller. He is both the father of *rhum agricole* – rum made from pressed sugar-cane juice – and the bogeyman who would steal naughty children.

Labat had seen how *rhum* could pay for a plantation's costs, but it was to no avail. The edict of 1713, which banned the importation of rum and molasses into France, put paid to any serious distilling. Instead of making *rhum*, planters sold their molasses to American distillers.

It took the British occupation during the Seven Years War for more advanced rum-making to be introduced to Martinique and Guadeloupe. By the end of the eighteenth century, there were 215 distilleries in Martinique, 128 in Guadeloupe, and 182 on St Domingue (which would later become Haiti). The smaller islands were already more oriented toward rum, while St Domingue concentrated on sugar. All that ended with the Haitian Revolution, the vicious Napoleonic backlash, civil war, mass white flight, and the collapse of its sugar industry. Haiti could have rivalled Cuba. Instead, it was shunned.

Emancipation finally took place in 1848, raising the costs of sugar production, while the development of beet sugar in France also impacted on the industry in Martinique and Guadeloupe. The islands drifted into a state of torpor, evocatively described in 1887

Martinique *rhum* reached its height of popularity in the late nineteenth century.

HISTORY

by American writer Lafcadio Hearn, in whose prose are echoes of the same romantic impulse that helped drive people in England north to the wild Scottish wastes. The remote was not to be feared, Hearn argued. The tropics were a languid, decadent paradise.

In Martinique he encountered an island where white settlers were leaving "at a rate that almost staggers credibility" and a *rhum* industry fighting a seemingly losing battle. *Rhum* was, however, still being drunk. Hearn writes: "the *mabiyage*... a popular morning drink among the poorer classes... is made with a little white rum and a bottle of the bitter native root beer called *mabi*. It is not until just before the midday meal that one can venture to take a serious stimulant – yon ti ponch – rum and water sweetened with plenty of sugar, or sugar syrup."

Rhum was saved by the introduction of new types of stills. Initially, distillers used the Derosne still, developed in France for distilling sugar beet. This was then adapted to suit molasses, with enrichment trays being installed at the top of the still. This became known as the Créole column (*see* p.50) and continues to be used.

Outbreaks in France of the oïdium mould in the 1850s and phylloxera vine louse in the 1870s devastated wine and brandy production, resulting in drinkers looking to the Caribbean. By 1896, France was

Sloppy Joe's in Havana, Cuba; the bar that never closed.

The island of Cuba offered hedonistic delights for Americans.

importing 28,640,367 litres (6.3 million gallons) of rum and Martinique was the Caribbean's top exporter, with Saint-Pierre the world's rum capital.

All of that changed in 1902, when Mont Pelée erupted, killing 40,000 people and destroying Saint-Pierre. What was left of the sugar industry went into retreat, major distillers closed, and smaller estates began to specialize in a new style, made from cane juice, which they called *rhum agricole*. The rum style map was complete.

Cuban Rum and Prohibition

As Martinique recovered from the eruption, Cuba was continuing its rise in importance and had become America's favoured producer. The main boost for Cuban rum – and Bacardí especially – came with Prohibition. Even if America was, ostensibly, dry, the "Great Experiment" would prove to be the boost that a still-struggling rum category needed.

Thirsty Americans headed straight to Havana, where they stayed at new, American-owned, hotels: the Sevilla-Biltmore, the Plaza, the Bristol, the Miramar, and countless others. Bartenders arrived: Eddie Woekle, Pete Economides, Vic Lavsa, and George Harris joining existing residents such as Galician immigrant José Abeal Otero, whose Sloppy Joe's bar was open for 24 hours a day. At the top of the tree was the Floridita, overseen by Constante Ribalagua.

Prohibition changed rum-drinking. It created a new age of cocktails: the Mulata, Constante's refined Daiquiris, El Presidente, and many more. Modern rum was forged here, the drinks shaken to the clavé rhythm.

The Affordable Playground

Havana was a steamy, erotically charged, crucible of creativity. As Louis A. Pérez writes in *On Becoming Cuban*, "[Cuba] was constructed intrinsically as a place to flaunt conventions, to indulge unabashedly in fun and frolics in bars and brothels, at the racetrack and the roulette table, to experiment with forbidden alcohol, drugs, and sex." It was an affordable playground, with 7,000 bars, a racetrack, golf courses, boxing rings, an amusement park, nightclubs, cabarets, and, in 1928, the Gran Casino Nacional. It was louche, but not too dangerous and, as English was spoken wherever the tourists would go, familiar. Cuba was a projection of America's base desires.

Even after the repeal of Prohibition, travel remained cheap, with gambling taking over from booze as the American Mob sought out a new source of income. Chief among them was Meyer Lansky, who had first come to Cuba during Prohibition to secure a supply of molasses to distil back home. Even after Prohibition was repealed, Lansky and some associates were behind the Molaska Corporation, which operated huge illegal distilleries in Cleveland, Ohio; Buffalo, New York; Chicago, Illinois; Zanesville, Ohio; and Elizabeth, New Jersey.

In 1934, Lansky flew to Cuba with a suitcase full of money and did a deal with dictator Fulgencio Batista. Batista got $3–5 million (£2.2–3.6 million) a year, and the Mafia got the monopoly on casinos. At least 178,000 Americans headed to Havana in 1937. That's a lot of rum drinks.

The lure of Cuba continued after the Second World War, with tourists heading to the island and rum being exported, along with "Cuban" dance crazes like the mambo and rumba. It could never end... could it?

MEANWHILE, IN THE BRITISH CARIBBEAN...

A virtual collapse in the British Caribbean sugar trade in 1901 focused estate owners' minds on rum. Science was now at the disposal of rum-makers, and Jamaica acted, creating the Sugar Experiment Station in 1905, run by H.H. Cousins and building on the work of Percival Greg and Charles Allan on fermentation and cultured yeasts.

The German market had also opened up, and with it, a new style. Germany had developed a taste for heavy rums, but in 1889 its government raised the duty on Jamaican rum. The response, proposed by Finke & Co. of Kingston and Bremen and assisted by the Experiment Station, was to create a new style of super-concentrated, high-ester (Continental Flavoured) rum. This could be blended and then diluted with neutral spirit in Germany to achieve (roughly) the same result – but with less tax.

In 1907, Jamaica's 110 estates were producing three grades of rum, showing the taste preferences for its three main markets. "Local trade quality" was

Jamaica's Appleton Estate continued producing through rum's hard times.

a quick-maturing, light rum for the domestic drinker. Fruity, heavy, pot-still, dunder-influenced "Home Trade Quality" went to the UK, and "Export Trade Quality" (Continental Flavoured) was destined for Europe.

Despite this, not all Jamaica's estates would survive. The twentieth century was, as for all rum-producing countries, one of consolidation. By 1948, there were 25 distilleries in Jamaica and outside investment was coming in, most notably from Canadian distiller Seagram, which needed a supply for its Captain Morgan brand and came to own Jamaica's Long Pond Estates and plants in Puerto Rico, Mexico, Venezuela, Brazil, and Hawaii.

Guyana was taking a different tack. The major supplier to Britain concentrated on short-fermented, pot- and column-still rums; there simply wasn't the vat capacity to run extended ferments. Though consolidation happened, rum sales were healthy. Barbados, however, was struggling. In 1906, distillers were barred for selling direct and all trade went through local merchants such as Alleyne Arthur, Martin Doorly, R.L. Seale, and Hanschell Innis. The trade shrank and became concentrated on the domestic market.

HISTORY

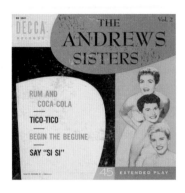

The Andrews Sisters song "Rum and Coca-Cola" immortalized the Cuba Libre in the 1940s.

RUM AND COCA-COLA

For the rest of the Spanish-speaking Caribbean, the task was simple. Catch up with Cuba and exploit the ever lightening US palate. Puerto Rico was well set until 1917, when, along with signing the Jones Act that extended US citizenship to its residents, citizens also – bizarrely – voted in favour of Prohibition, killing a rum industry worth $70m a year. It would take until 1935 for rum distilling to restart. Even then, most "Puerto Rican" rums were blends of other islands' rums with added flavours, wine, sugar, molasses, prune, and other fruit juices.

This wasn't unusual. Post-Prohibition and throughout World War II, many distillers were simply filling pipelines. Rum had become a low-priced alcoholic hit. A fascinating report on rum styles from 1937, written by Peter Valaer from the US Treasury, shows that light, column-still Cuban rums were "fruity, or slightly like the taste of molasses", while the four St Croix distilleries were using cane juice, a mix of inoculated yeast, or wild ferments; and pots and columns.

Puerto Rico's rums were mostly molasses-based, while "quick ageing" techniques such as "[ageing] in a white oak barrel and treated with calcium permanganate and hydrogen peroxide" were used. While it would seem that Jamaica's quality was untouched, all of the rums profiled by Valaer were coloured or adulterated. According to him, Demerara rum producers "added French plums, Valencia raisins, spices, and other flavoring ingredients" before colouring and ageing. He also claimed that "some... have a unique custom of placing chunks of raw meat in the casks to assist in aging, to absorb certain impurities, and to add a certain distinctive character".

Barbados was now using modern column stills and adding flavour through the use of "sherry, Madeira or other wines, often spirits of niter [sic], bitter almonds, and raisins".

Despite this, rum had a good war. The American whiskey industry had closed down (again) in order to make industrial alcohol for the war effort, and people were thirsty. There were also large bases on the Caribbean islands (notably Trinidad) and in World War II where the troops went, so did Coca-Cola, which ensured that there was a bottling plant close by any major military installation.

The mix first made famous during the Cuban War of Independence, the Cuba Libre, was now simply "Rum and Coca-Cola", immortalized in song by the Andrews Sisters: the biggest hit about prostitution until "Ruby, Don't Take Your Love to Town".

Postwar, things changed fundamentally in Puerto Rico with the creation of the Rum Pilot Plant under chief chemist Rafael Arroyo, whose papers remain essential reading for rum geeks – especially his analysis on how to make heavy rum. His conclusion? Don't take it from the first column: "... a carelessly distilled light rum is not a first-class, genuine, heavy rum". In 1943, Puerto Rico exported more rum than Cuba to the US, where it was now being drunk in a very different way.

Enter Tiki

In one corner, an extravagantly named former teenage rum runner Ernest Raymond Beaumont Gantt, aka Donn Beach, aka Don the Beachcomber). In the other, Vic Bergeron, aka Trader Vic, a wooden-legged salesman/ hustler from Oakland.

In 1934, Donn Beach opened the Polynesian themed Don the Beachcomber restaurant in LA. Decked out in his own Polynesian props, its speciality was wild new rum drinks. He called them *tiki*, and tiki was to become a major postwar craze in America. Tiki was fun, vibrant, exotic. It was everything that the grey flannel life of the 1950s wasn't. It also allowed rum to do what no other spirit does as well: make people smile.

What was forgotten until American tiki historian Jeff "Beachbum" Berry (whose work has helped immensely with this section) revealed the truth, was that for all the kitsch, these original drinks were good. OK, they were somewhat rococo, but they worked. Donn and Vic knew their rums.

Donn loved Jamaican pot-still rum. He blended rums in his drinks; he blended juices and spices; he took punch principles and turned the volume up to 11. Vic had been quick to spot Donn's success, and in 1937 offered to partner him. Donn rebuffed him, so Vic, always the smarter entrepreneur, headed to Cuba to learn from Constantino Ribalaigua Vert, aka Constante. In 1938, the

Trader Vic's themed bar/restaurants helped to introduce tiki culture.

first Trader Vic's opened, complete with Vic's Daiquiri riffs. In 1944, he went one step further, inventing the Mai Tai. At its peak, his chain had 20 restaurants.

Things went less well for Donn, who lost half of his empire when he divorced his wife. He left for Oahu to continue his tiki life.

Rum was now miles away from its origins. For an American drinker, it meant potent fruity drinks. In the Caribbean it was white, drunk with water, or a simple mixer. In Britain, it was a heavy, dark spirit, taken neat or with toxic syrups.

There was another drink, and it was ubiquitous. It was called Bacardí, but by the 1960s, certainly in the UK, it wasn't thought of as a rum. It was simply "Bacardí", and it was a glorious success for the firm which had managed to do what no other brand had: transcend the category.

The Business of Rum

Rum's story in the latter part of the twentieth century becomes a tedious litany of consolidation, nationalization, and mergers. As the tectonic plates of alcohol shifted, so rum became the collateral part of multinational deals. There was little talk of the spirit. By the 1970s, sales were on the slide. In Cuba, meanwhile, the industry had been nationalized and export ceased. Elsewhere, it was kept afloat by tax breaks, such as those given to Puerto Rico and the US Virgin Islands by the US, or duty-free quotas handed out to the Caribbean producers by the EU. The quota system saved rum production, but it did little to help develop brands. Rum was bulk, or it was Bacardí.

In 1997, the quotas and preferential tariffs for UK/French rums were dropped (although the US tax breaks for Puerto Rico and the US Virgin Islands stayed). It took a $70 million (£50.5 million) EU package in 2001 to help stabilize the market, with grants given to assist in promotion, production, and the development of distiller-owned brands.

Another battle was underway. In 1976, the Cuban government took the lapsed Havana Club trademark and began to market the brand once more. In 1993, Pernod-Ricard entered into a joint-venture agreement giving it global marketing rights. The possibility of a global, premium, aged rum brand had become a reality.

The Mojito; the twenty-first century's go-to rum drink.

Bacardí – which had its Cuban assets seized in 1960 – saw things differently. Thus began a prolonged legal battle over the rights to the Havana Club trademark, which is still continuing. Absurdly, despite the Cuban embargo being on the verge of finally being lifted, Bacardí is still blocking the sale of Havana Club in the US.

RENAISSANCE

It was 1993. The bar was still a building site, offering glimpses of what it would become – the Atlantic Bar & Grill, shrine of a new cocktail age for London. In a side room we gathered. Oliver Peyton, Dick Bradsell, wine writers, myself, and a small selection of something new: "gold" rum.

There were pioneers like Mount Gay, Appleton, and Cockspur and the recently launched El Dorado 15-year-old: for me, the brand that was the first to create a premium rum category and one whose producer proudly stated "We make this." The mind-forged manacles of colonial thinking were loosening.

We left enthused, converts to a cause, convinced that rum's accession to centre stage was inevitable. It took longer than rum-lovers around the world anticipated, but every year since has seen more rum bars open, while tiki's recent revival has increased the momentum. Globetrotters like Ian Burrell and Velier's Luca Gargano tirelessly spread the word. Bacardí's re-establishing of its rum credentials gave things an additional boost, as did Havana Club's Cultura initiative. The West Indies Rum and Spirits Producers' Association (WIRSPA) generic campaign, rum festivals, blogs, books, and magazines all added their voice.

Most importantly, the producers kept faith. They talked of history and place. They spoke of Guyana or Jamaica, of Cuba or Guatemala, of Barbados or Nicaragua. They spoke of flavour and heritage. They spoke of belonging.

Rum had come home.

The people who made it were, finally, in charge.

PRODUCTION

Legally, rum can only be made by distilling the products of cane-sugar manufacture – molasses, syrup or fresh cane juice. It must be distilled no higher than 96 per cent for the EU and 95 per cent for the US. In the EU it cannot be bottled at lower than 37.5 per cent ABV. In the US the minimum strength is 40 per cent. The EU does not allow flavouring to be added to rum, but the US does.

Other than that, on the face of it, rum's pretty simple. Take a sugar solution, add yeast, ferment it, then distill. At that basic level that's all every rum distiller does. What makes rum fascinating and downright mysterious is how each one approaches these straightforward principles in such different ways.

Every point in this decision-making process influences a rum's final flavour. Technology is used, but not at the expense of artistry. Terroir exerts itself in a multiplicity of ways, from a cultural aspect to a direct manifestation of soil, wind, and air.

Rum-makers, be they distillers or blenders, are guardians of style, holders of history, custodians.

This is their story.

This is their world.

Some sugar cane is still hand cut.

CANE

It is fair to say that you don't get a lot of practice with a machete in Glasgow. If you try, you tend to be arrested. Still, being given one in a plantation gives me an idea of the sheer bloody physicality of cutting cane. This is not romantic in any way. It's a relentless, back-breaking slog that continues all day, every day during the dry season.

Not all cane is hand cut today. While its proponents claim it's the best way to capture the high concentration of sucrose at the base of the stalk, most cane is machine cut, allowing harvesting to take place 24:7.

The canes I'm cutting – varieties chosen to suit soil and climatic conditions – were planted a year ago and have grown to between three and five metres (10–16 feet). The fields have been burned to sanitize the soil, remove the trash (dead leaves), and scorch the cane, preventing water loss when it's cut.

As soon as that happens its composition starts to change. The invertase enzyme begins to turn sucrose into glucose and fructose, while dextran builds up. Because all these compounds make it more difficult to crystallize sugar, the cane has to be at the mill within 24 hours.

Most producers believe that the variety of cane does not matter if you are making rum from molasses. An exception is Appleton Estate, where master blender Joy Spence claims that the varieties they use help produce a fruity and slight buttery note in the rum.

It will, however, have a significant effect for rum distillers who use cane juice (*see* pp.29–31).

Sugar Production

I never tire of visiting sugar mills. The trucks rumbling in, piled high with teetering piles of cut cane; the massive grabbers picking up canes, feeding the insatiable maw of the mill. The smell is all-pervading: a weird, heady mix of sweet and sour, moist earth and vegetation.

Here the canes are chopped and crushed to release their juice, which is then clarified with a mix of lime to neutralize its pH. The resulting mud can be racked off and used as fertilizer.

The juice's pH is upped and evaporated into a syrup, then concentrated in vacuum pans into a

PRODUCTION

supersaturated state. Tiny sugar crystals are seeded in, prompting larger ones to form, which then can be centrifuged away. This is then repeated twice. One on side is raw sugar, on the other molasses.

Molasses and Cane-syrup Rums

By this stage, the rum-maker already has three options for a base material. Cane juice can be used (*see rhum agricole*, p.48), as can cane syrup (the choice for many Latin American rums) the viscous, sweet product extracted after the first crystallization. This costs twice as much as the heavy, thick molasses with its bittersweet, smoky tang of iron and blood, the most common base.

Molasses

The consolidation of the global sugar industry means there is also a bulk trade in molasses, which comes predominantly from Brazil, Guyana, and Venezuela. This will be the source of raw material for distillers in countries where there is no (or little) sugar being made. Each distiller has precise specifications for sugar content, ash, gum, pH, and acidity to ensure his or her rum's flavour profile is consistent.

The knock-on effect of greater efficiency in sugar-making is lower sucrose levels in molasses and a rise

It is vital to get the sugar cane to the mill as quickly as possible.

Molasses arriving at the Brugal Distillery, Dominican Republic.

in its ash content. The latter can cause issues in the still, while a lower sugar content has a knock-on effect on yield and cost, meaning producers require more molasses to produce the same amount of rum.

Yeast

Yeast can either be seen as the organism that converts sugar to alcohol, or an active participant in flavour creation. Today, wild yeast ferments (using yeasts naturally occurring in the local environment) are rare, but they do still occur. Some distillers use standard dried commercial yeast, while others have developed their own strains to produce specific flavours.

Put yeast into a sugary environment, and it goes crazy, emitting heat, and CO_2 and, most importantly, turning those sugars into alcohol. It is also temperature-sensitive, needing heat to get started, but unable to survive above 35°C (95°F). Maintaining that upper limit isn't an issue in cool climates, but it becomes a problem when the ambient temperature is between 25°C and 32°C (77°F and 89.6°F), as it is in rum-making regions. Temperature control is therefore required to keep the yeast working, prevent spoilage, and ensure that full conversion takes place.

Molasses is one and a half times denser than water, meaning that no yeast could work through its thick, cloying depths. Prior to fermentation, therefore, it is diluted with water, the level of which varies, depending on the desired flavour profile.

Yeast also requires nitrogen in order to work, and as molasses is low in nitrogen the level will be adjusted pre-ferment, most commonly with the addition of ammonium sulphate or ammonium phosphate. As yeast also benefits from a slightly acidic environment, the pH needs to be lowered to around 5.5 to 5.8. This in turn helps in the creation of fruity esters.

Fermentation

"Longer time for more flavour." *Mark Middleton, distillery manager, Hampden Estate, Jamaica*
Distillation is about the separation of alcohol from water, concentration of flavours, then selection of the flavours you want. Where do these flavours come from? Fermentation.

PRODUCTION

Molasses contains 81 aromatic compounds that react with the yeast, which itself adds flavour. These then interact over time to create more flavours. The length of the ferment therefore has a significant impact. The longer it runs, the higher the acidity of the wash, and the more esters you make.

A light rum needs a rapid fermentation – between 24 and 48 hours – to get the correct strength and flavours required. In general, heavier rums need longer ferments. Here, after the sugar is converted into alcohol, the wash is kept in the fermenter, where lactobacillus begins to work, helping to create those esters. Control is needed, whatever method you use.

Temperature control allows distillers to play around with these basic rules. Zacapa, for example, uses a 100-hour temperature-controlled ferment, while on the island of Marie-Galante, the Rhum Rhum brand distilled by Gianni Capovilla at Bielle is fermented for five days in a temperature-controlled environment. At Foursquare Distillery in Barbados, Richard Seale slowly feeds in his molasses-and-water mix into the fermenter to help create more complex aromas.

The length of time spent fermenting will have a direct effect on a rum's flavour.

DUNDER AND MUCK PITS

Dunder (see p.24) is now in the hands of specialists such as Jamaica's Hampden Estate and is used mainly for higher-ester rums. The Hampden approach uses a combination of molasses, water, dunder, and some liquid taken from the distillery's muck pits. This is made up of the residue gathered from all the tanks in the distillery at the end of the distilling season, which is dumped into a trench that is never emptied. Naseberries (sapodilla), jackfruit, and banana are also added to help boost nitrogen levels. It's the dunder and the muck pit that bring the funk.

Jamaica, however, is the fermentation specialist. Here, fermentations can be as short as 30 hours, or as long as Hampden Estate's 21 days. The rums that are produced are then graded by their ester levels. The well-named Common Cleans are between 80–150 esters and are made by short ferment. A fruity, raisin-accented Plummer (between 150–200 esters) might have a two-day ferment, while the oily, fruity funkiness of a Wedderburn (200+ esters) comes from a longer fermentation – and the option of the use of dunder (see box, p.24). Continental Flavoured rums (between 700–1,400 esters) have massively extended fermentations, smell of acetone, and are used mainly for flavouring purposes.

DISTILLATION

The distiller has created a wash of between 4% ABV and 9% ABV, with all of the flavours held in a dilute solution. Because alcohol boils at a lower temperature than water, if you heat this in an enclosed vessel, the alcohol will be driven off in preference to the water, thereby increasing the strength and concentrating the flavours.

The aroma-rich vapours rise up the still and are then converted back into liquid in a condensing system, which is usually a bundle of copper tubes filled with cold water. The more you make the vapour work, the higher its strength, and the lighter in flavour the spirit will be. Conversely, the shorter that journey, the heavier in flavour it will be.

What's happening is reflux. When the vapour meets a slightly cooler part of the still, its heavier elements revert into liquid and are redistilled: micro-distillations within a greater whole. Reflux breaks down the vapour stream into increasingly small fragments, thereby revealing more of its complexities.

Stills are usually made of copper, and as copper holds on to heavy elements such as sulphur, the longer the dialogue between vapour and copper, the lighter the spirit will be. So the slower the distillation – or the taller the still – the longer this conversation is.

The shape and size of the still and the speed of the distillation will also have a direct impact on the final spirit character.

Pot-still Distillation

The traditional method of maturing heavy rum is still practised by many distillers.

Traditional Pot Stills

A pot still is effectively a giant kettle. The first distillation will give a spirit, called low wines, of around 24% ABV. This needs to be redistilled to refine the flavours and increase the strength. Here, the volatile elements in the first portion, known as the heads, are separated from the body of the spirit (the "heart"). Toward the end of the distillation, the flavours become increasingly oily. As these are undesirable, this portion, called the tails or feints, is also separated. The end result is a spirit with a strength of around 65–72% ABV. The heads and tails are redistilled with the next batch of low wines.

There is a range of flavours within the heart for a distiller to choose from. A lighter, more fragrant rum will be collected from the earlier part of the run, a richer one from the distillate coming across later. These heavier elements give mouthfeel and backbone to a rum that is going to be aged.

Pot Plus Retort

Rum distillers also use a system in which copper vessels called retorts are placed between the pot and the condenser. The pot is filled with wash, while the retorts contain the high wines and low wines from the previous distillation. This allows triple distillation to take place in one pass of the still.

The wash is heated and the alcohol vapour (at around 30% ABV) is carried into the first retort, which holds low wines. This then boils, liberating all of its aromas, and the vapour (now 60% ABV), goes into the second retort, where the same thing happens. The vapour, (now at 90% ABV), is then condensed.

As the liquid flows into the receiver it's split into four parts: heads, then rum (with an average strength of 86% ABV); next, a cut is then made to high wines, with an average of 75% ABV, then a final cut to low wines, with an average of 30% ABV. The last two are used in the retorts for the next run, acting like a chicken carcass that flavours a stock. By adjusting their strength, the distiller can create different end results.

The hybrid still at St Nicholas Abbey distillery, Barbados.

The pot and retort system at Foursquare distillery, Barbados.

If making higher-ester rums, the distillation will be run through the whole system a second time to concentrate the aromas further.

Wooden Pots

Demerara Distillers' Diamond Distillery has two wooden pot stills. The first, the Versailles still, is made up of a pot of greenheart wood with a copper neck that plunges into a retort, and from there into a small rectifying column (which helps add more reflux), and then to the condensing system.

The second of the wooden pot stills, the Port Mourant still, has two wooden pots. Both are filled with wash, then the heat is turned up fully on the first to drive off the alcohol. This vapour surges into the base of the second pot, boiling it, and this stream then runs into a retort and rectifier.

As there's not a huge amount of copper here, the resulting spirit is heavy. That from Versailles (VSG) is beefy and rich. Port Mourant (also known as Port Morant, or PM) has a hint of oiliness, with added black banana, and overripe fruit. Both need long ageing and add substance to blends.

PRODUCTION

Column Stills

The invention of the continuous or column still in the nineteenth century ushered in a new era for rum (*see* p.23). For the first time a "light" rum could be made. Today's rum distillers use a number of different designs of column still.

Coffey Stills

Designers like Aeneas Coffey were looking for efficiency in distillation, with designs that allowed a "continuous" process. In a Coffey still's case, this is achieved by linking together two columns: the analyser and the rectifier, both divided internally by perforated horizontal plates into a series of chambers.

The wash is carried down the rectifying column in a coiled pipe, which rises to the top of the analysing column, where it is sprayed over the top plate. It then flows downwards through a series of channels. Live

The five-column still set up at Angostura distillery, Trinidad.

steam is being injected into the column's base, passing through the perforations and stripping off the alcohol from the descending wash.

This vapour rises and is gathered in another pipe, which feeds into the base of the rectifying column, then releases the vapour to continue its upward journey. As it progresses, it hits the slightly cooler top of each chamber, thus causing heavier elements to reflux out.

It behaves like a cross-country runner starting a winter race, wrapped up in heavy clothes. As she heats up, so various layers (flavours) begin to be shed.

Because the columns are immensely tall, only the lightest compounds can reach the collecting plate, where they are diverted into the condenser.

Multiple Columns

The more columns in your set-up, the greater control you have over the distillate – and the lighter the rum will be. These configurations have more bells and whistles, allowing different alcohols to be collected, redistilled, or removed.

A distiller can make a huge range of different marks with this system. Demerara Distillers Limited (DDL), for example, makes nine different marks from its four-column Savalle stills. In Jamaica, Wray & Nephew makes an unspecified number of marks on its three-column set-up. Five-column stills are used by distillers such as Bacardí, Cruzan, and Angostura.

Bacardí's starts with a "beer column", which strips off the alcohol and yields the firm's *aguardiente* (80% ABV). After being run through a further three columns, all the unwanted elements have been removed and Bacardí has its *redistilado* (95% ABV). The fifth column is used for redistilling elements from the four other columns. All of this is done under vacuum, which lowers the boiling points and makes the process more energy efficient (Foursquare also has a vacuum column still).

It takes great skill to make a "heavy" rum from a stripping column, and it requires a lot of sulphur-removing copper. There is a risk of heavy fusel oils being retained. The resulting rum might be the same strength as a pot-still rum, but because it isn't fermented and distilled in the same fashion, its flavour and palate weight will be different.

Rhum agricole is distilled in a single column still, like the one shown here.

Hybrid Stills

These stills combine a pot with a rectifying column (sometimes contained in the neck of the still). Bacardí's original still was an early version of one of these hybrids. Modern examples are used at Saint Lucia Distillers, and St Nicholas Abbey in Barbados.

Filtration

Typically, light or "extra-light" rum will be passed through charcoal to remove any aggressive elements. This can either be bottled, or matured. In Bacardí's case, both its marks are filtered before maturation. After that, like a number of white rums, a second filtration takes place to remove colour.

OTHER RUM STYLES

Rhum Agricole

Rhum agricole is made from fresh sugar cane juice. (If molasses is used, the spirit is called *rhum industrielle*.) *Agricole* has a physical link between the spirit and the land. If you're using cane juice, you need to process it as quickly as possible. This means having the mill and the distillery on the same site; often the cane is grown on the surrounding estate.

Juice also allows you to notice the influence of sugar cane – even the variety: how the influence of soils impact on the *rhum*; or the impact of climate on a *rhum* made from a distillery cooled by Atlantic breezes compared to one on the hotter Caribbean coast. As distillation is about a complex interplay between different factors, it is hard to point at a single aspect and say, "This is what makes the difference." All of these factors – cane, soil, climate – will impact in some way.

Taste a range of vintages and you can discern the differences between the climatic conditions of each year. The punch of a 50% ABV *rhum agricole blanc* takes you straight to the island's earth, a vegetal note mixed with that cane-juice aroma of flowers and light fruit you smelled in the distillery.

Rhum agricole in many ways is one step closer to wine. You need to use what the vintage gives you. These distilleries are châteaux of *rhum*.

Production

On arrival, the sugar cane is crushed, the juice diverted one way, and the fibrous mass (*bagasse*) used as fuel for the boilers. The juice tends to be fermented rapidly in open fermenters with dried yeasts. Some distilleries complete this within a day, others run the ferment over two. The maximum permitted in Martinique is 72 hours.

Distillation takes place in single column stills separated into 20 to 30 plates. The fermented wash (*vesou*) is passed through two preheaters, then directed into the middle of the column, while live steam is pumped in from the bottom, stripping off the alcohol from the *vesou* as it flows downward over the plates.

This vapour stream rises above the vesou's entry point into the still's enrichment section, where reflux happens. It then passes through a pipe running through the preheaters (providing the source of heat) to the condenser. Any refluxed alcohol collected prior to condensing is fed back into the column. The result is a lower-strength spirit. Under appellation law in Martinique, the distillate must be between 65% and 75% ABV.

The differences between *rhum agricoles* can thus come from terroir, cane type, the length of the ferment, the strength of the *vesou*, and the physical make-up of the column: its height, copper content, and how the chambers are separated.

Gauging the weight and volume of a cask at Clément distillery, Martinique.

The old "Creole-style" still is the simplest design and produces more spiciness in the distillate; Savalle and Barbet designs make the vapour work harder, increasing reflux. Savalle stills tend to make a more floral style, while those from Barbet set-ups are often weightier and more vegetal.

This green, vegetal note comes primarily from fermentation and distillation. This can be shown when you compare an *agricole* with a pot-still, cane-juice rum such as Rhum Rhum, which has had a long fermentation. Distilled to a higher strength, it is clean, floral, and fruity, with no vegetal notes.

Clairin

It was commonly thought that Haiti only had one distillery: the legendary Barbancourt. The reason? It was available worldwide and few people visited that sadly troubled country. Then rumours began to circulate that there were more – a lot more. In fact, there were hundreds of stills in Haiti. What they produced however wasn't rum, or *rhum*. It was *clairin*.

Clairin is to rum what mezcal is to tequila and should be approached in the same way. This is a small-scale, artisanal, agricultural spirit, made in a traditional rather than an advanced technological manner. It has the soil of Haiti in its blood.

Cachaça

Brazil is the world's largest sugar producer, and its sugar cane also provides industrial ethanol, molasses, and cachaça: 1.5 billion litres (329,953,872 gallons) of it a year, this quantity made by an estimated 30,000 registered producers.

More than 90 per cent of production is concentrated in the hands of a few large producers, who make "industrial" cachaça. A vastly larger number of smaller producers make "artisanal" cachaça with Minas Gerais having the highest concentration at around 8,500 of them.

By law, cachaça is a spirit made from sugar-cane juice (*garapa*); in Brazil, it must be distilled to no less than 38% ABV and no more than 48% ABV, to which no more than six grams per litre (1/5oz per 1¾ pints) of sugar can be added. Industrial producers will source cane from a wide variety of plantations. Artisanal

A triple pot set up for cachaça distillation.

producers tend to own their own cane fields, giving them greater control over the base ingredient.

After clarification of the juice, fermentation can start. Some industrial producers utilize techniques from the fuel-ethanol business and can complete fermentation in 10 hours. Artisanal producers will run their fermentation for longer. In their case, the yeasts on the cane provide the fermenting agent.

To this, some add a "starter" made of toasted corn/flour/bran/soy bean/rice that's mixed with *garapa* to kick-start the ferment. In many cases, this is only used at the start of the distilling season, with subsequent ferments helped along by the addition of some of the ferment from an older batch. This technique, which is akin to that used in making sourdough bread, is commonplace, but with so many distillers there are an infinite number of variations.

Fermentation of artisanal cachaça is gentle, taking between 24 and 36 hours to complete the conversion to fruity, mild *vinho* (literally "wine"). The yeast is then racked off and re-enters the system, while the *vinho* is distilled.

Industrial cachaça is a column-still distillate, while artisanal is made in copper pot stills where the heart

Cachaça can be mellowed and aged in a wide variety of woods.

of the distillate is separated from the heads and tails. There are a huge number of different styles of these: simple pots and condensers, some with plates in the neck, others with two linked pots, both with plates in the neck.

Industrial producers sell "fresh" cachaça, while artisanal makers will always give their spirit a period of mellowing in casks of no larger than 700 litres (154 gallons). While ex-bourbon and ex-Cognac barrels are common, a huge variety of native Brazilian woods are also utilized: amendoim (amaranth: *Pterogyne nitens*) and tauari (from the *Couratari* genus) – both confusingly called "Brazilian oak"; ipê (*Tabebuia*), imburana (*Amburana cearensis*), freijo (*Cordia goeldiana*), and jequitibá (*Cariniana legalis*).

Some, such as the first pair, are neutral in their character, making them ideal for mellowing. Other types have more colour and aromatic qualities and are used for maturation.

Is this rum? No, it's cachaça!

Arrack

The term "arak" is a catch-all term for a distillate. In Sri Lanka and Goa it means a spirit made from the coconut palm. In Lebanon it is a grape-must spirit, while in Mongolia, *arkhi* is made from fermented mare's milk.

Batavian Arrack, the go-to base for classic rum punches is now available once more.

In Java, however – where an extra r and c are added – the term refers to a cane spirit.

Batavian arrack was, in its time, the most highly prized cane spirit for use in punches (*see* p.19). Produced in Java, it was shipped mainly to The Netherlands, from whence it was re-exported. The firm of E&A Scheer (*see* pp.55–6) was primarily an arrack importer before it became a wider rum specialist. It still trades with Java and is the source of Batavia-Arrack van Oosten. The spirit is also used by the flavour industry and in Swedish *punsch* (a blend of Batavian arrack, rum, sugar, and spices).

The spirit is produced from molasses, which is fermented with the addition of a starter made of red rice cakes. Distillation takes place in pot stills, after which the arrack spends time mellowing in teak vats before being shipped. Further maturation can take place in Europe.

MATURATION

Rum was one of the world's first deliberately aged spirits. Legally, some countries insist that rum has to be aged before it can even be called rum; in Cuba, for example, the spirit has to spend two years in cask before it can be called rum.

An understanding of oak is the area that is benefiting rum the most. After all, if up to 70 per cent of a cask-matured rum's character comes from the interaction between oak and spirit, then the nature of the oak, its quality, and character are vital.

Oak is used because although it is watertight, it also breathes. Liquid stays in but oxygen (and alcohol vapour) can pass through. It is durable, so casks last a long time. It is also easily coopered. Most importantly, it has flavour.

Casks from the bourbon industry are by far the most widely used. These are made from American oak (*Quercus alba*) and add notes of vanilla, coconut, chocolate, and sweet spice to the spirit. Producers of *rhum agricole* have traditionally used ex-Cognac casks made from "French oak" (*Quercus sessiliflora*), which add savoury spiciness, vanilla, and grip. Less common are ex-Sherry casks made from "Spanish oak" (*Quercus robur*), which add notes of cloves, resin, dried fruits, and tannin. Distillers such as Brugal and Foursquare have had success with these, while the latter has also begun to explore other fortified wine (and regular wine) cask types.

Oak should participate in the slow accretion of flavour. It gives colour; it mellows; it adds character.

Mechanisms of Cask-ageing

The first thing that happens when rum is put into a cask is the removal of aggressive elements, either through evaporation or by the layer of char on the inside of the cask, which acts like the charcoal in a cooker hood.

The inside of the cask will also have been toasted. This transforms aromatic compounds in the wood, creating flavours such as coconut, vanilla, spice, and cacao. It also creates tannins, giving colour and grip. All of these seep into the rum and interact with the flavours created in distillation, increasing the spirit's complexity.

The fresher the cask, the more impact it will have. A three-year-old rum put into a fresh cask will show more oak influence than the same rum put into a cask that has been filled a number of times. Refills still have flavour, but are more discreet. What needs to be avoided is a cask that has been exhausted and has turned into a container rather than a participant.

While a heavy pot-still rum needs time to fully mellow out, a light rum can get oaky too quickly. Having a deep understanding of these parameters is part of a rum blender's job.

Climate

As the rum matures, so the cask is breathing, inhaling oxygen (which helps change the aromas) and exhaling alcohol: the angels' – or, in the Caribbean, "duppies" – share. The hotter the climate, the more rapid this respiration is. Not only do you get quicker oxidation, but the volume of liquid is reduced and the interaction between rum and wood is speeded up.

A rum moves through its maturation cycle more rapidly in the Caribbean than in Europe. The same rums matured in Europe and the Caribbean are very different beasts after five years. The latter become more woody quicker than a cool-climate-aged spirit. Remember this when you look at an age statement on a rum's label.

PRODUCTION

Tropical ageing will have a marked effect on the flavour of a rum.

ZACAPA'S SISTEMA SOLERA

There are a number of different types of solera system. Guatemalan producer Zacapa's is undoubtedly the most complex, mixing static ageing, elevage and solera. The rum matures in a progression of different cask types: ex-bourbon, extra charred ex-oloroso, then ex-PX butts (and, in the case of XO, ex-Cognac), before entering a solera. In addition, each time the rum is moved to the next type of cask, a progressively older age profile of reserve rum is added to the blend while some of the blend being transferred is then blended back into the reserve.

Carsten Vlierboom, master blender at E&A Scheer, demonstrated this to me with a flight of rum from Jamaican-aged Worthy Park. The subtle, sweet fruitiness and light oak after one year in cask had turned into dry spice and raisin 12 months later. A year on, there were cedar notes and greater complexity. After four years, the wood was taking over.

If rums suck up wood quickly, the logical solution is to use a higher percentage of "refill" casks to give a gentler, more subtle result. There is a quality difference between a mature rum and one that's just dominated by the vanillin from a cask.

Another option is to locate your maturation cellars in a cool zone, an option taken by various Central and Latin American producers. Zacapa, for example, distils at 275 metres (902 feet), but its ageing cellars are in Quetzaltenango at 2,300 metres (7,546 feet).

Scheer matures its rums for a maximum of five years in the Caribbean before transferring them to the cooler conditions of Amsterdam or Liverpool, where the ingress of oak is dramatically reduced.

PRODUCTION

Understanding the different energies given to the maturing rum by climate, oak type, and cask activity is helping to widen the possibilities for rum.

BLENDING

Most rum is blended: a blend of casks, styles, distilleries, or countries. Blending gives volume, creates consistency, and establishes character. It is creative and dynamic.

If a distillery makes one style of rum, then the blender could use different ages of rums, or fills of cask to build in complexity. This would be the option taken by Cruzan in St Croix, for example. Havana Club takes a different path (*see* box, p.57).

Using different ages also helps build complexity. Young rums add vibrancy and freshness, older rums add depth. First-fill casks give a hit of vanilla, while refills allow the character of the rum to speak.

Other distillers may be producing different marks from pot stills, giving weight, and column stills, giving delicacy. These different characters, aged in different fills, can then be blended, which is the approach taken by Foursquare, Demerara Distillers, and Appleton Estate.

"We are always trying to find new flavour profiles," says Juan Piñera Guevara, Bacardí's master blender. "This could be through fermentation conditions, distillation, and casks; ages and fills; casking strength; distillates blended during ageing, etc." In other words, the blender is involved in the whole of the process.

To get an idea of the possibilities in blending, I visited Carsten Vlierboom, master blender at E&A Scheer in Amsterdam. Firstly, we looked at how to add character to a light (unaged) rum. Adding a small amount of Plummer gave raisin and punchiness, while a Wedderburn brought in tropical fruits, peels, and that Jamaican funk. A drop of Continental took things into pineapple territory. The blends seemed sweeter, but no sugar was being added. "You don't need to when you blend like this," he said.

He then showed me how different ages of cask-matured rums from one distillery could be blended to make a more complex whole, and how a rum which, on its own, would be too woody, can add structure to a blend.

Finally, we looked at how to make a multinational blend, with a selection of mature rums from Guatemala

Joy Spence of Jamaica's Appleton Estate; the first woman to break distilling's glass ceiling and become a master blender.

(coffee bean, savoury, oily), Barbados (vanilla, banana split), Nicaragua (chocolate and crispness), and more. Adding any of these, or combinations of these, to a light, aged base added depth, complexity, and aroma. The possibilities, even with this small selection, were huge.

Having created a flavour profile, the blenders must maintain it – even if volume increases, or a distillery stops producing. Ultimately, the buck stops with them.

Adjustment

Colouring rums is an old practice, and either spirit caramel or a molasses-base solution can be used. This is the standard practice for making the "Navy" style of rum. Here, the colouring will have an impact, not just on the rum's hue, but on its flavour, with a bitter, liquorice element being added. Cask-aged rums can also be tinted with spirit caramel. This is primarily to standardize any colour differences between batches.

Prior to bottling, some firms add sugar solution to their rum. This is currently the most vexed issue facing rum. Producers in Jamaica – which, like Barbados and Martinique, forbids sugar addition – think that declaring sugar levels should be mandatory.

An outright ban is extremely unlikely. No addition of sugar could, I feel, be declared on the label, much as some whisky labels declare that there's been no caramel tinting or chill-filtering. Better still, there should be a maximum amount of sugar permitted, as in cachaça or Cognac.

HAVANA CLUB'S "BASES"

Havana Club's single distillate is aged for two years in order to be classified as rum. It is then blended with high-strength sugar-cane alcohol. By differing the ratio between aguardiente and spirit, a wide selection of base rums (*bases frescas*) are made. A base with a ratio of 90 per cent *aguardiente* and 10 per cent high-strength rum will obviously taste different to one that is 10:90. These bases are then put into refill casks. Further blending takes place during maturation, with a percentage of each of the final blends being retained for further ageing as a base.

The result is around 20 bases, whose aromas range from geraniums to crème brûlée, pineapple to chocolate, jasmine, vetiver, and figs, but rarely any woodiness. The potential combinations are mind-boggling.

PRODUCTION

HOW TO USE THIS BOOK

You might well have flicked through the book and seen that there are 110 rums lurking around the corner. You may have even noticed that each of them appears to have been tried in different ways. The question "Why?" might have sprung to mind.

The answer is simple. I have this rum in front of me. What is the best way to maximize my enjoyment of it? If rum is so versatile, how does it play out when placed with a selection of the most popular mixers, or used in cocktails that offer no hiding place for the spirit?

The rums chosen include the major brands, other rums that are widely available, and some lesser-known ones that are top examples of a specific style. Space, sadly, meant that there was little room for cachaça and, for different reasons, there are no spiced brands. Hopefully the selection will help you to find new ways of looking at old favourites, reconsider rums you have bypassed, and discover some new friends.

SCORING SYSTEM

5* The best: this is a must-try. The rum and the mixer are in perfect harmony.

5 Superb. There's enhancement of the rum and a drink that is more than the sum of its parts.

4.5 Halfway between a superb drink and a great drink.

4 A great drink. The rum is opening out, thanks to the mixer.

3.5 Halfway between a great drink and a good drink.

3 Good. A sound and balanced drink. I'd be happy with one.

2.5 Halfway between a good drink and a drink that's just so-so.

2 Uh-oh. There's either no enhancement, or a clash between the elements.

1 Avoid.

N/A Some rums just are best on their own.

These scores are for the mixes and not for the rum, each of which was chosen for its inherent quality, so read the tasting note to glean how that manifests itself. Some are simply best left alone. Most only perform well with a few mixers; it is a rare beast indeed that works across the board.

THE LOWDOWN

This thing called rum comes in many guises, but some abiding principles apply whether you are tasting white, light, heavy, pot still, aged, or Navy. You are always looking for balance, complexity, and character. Are the sweet notes balanced by dry, or vice versa? Are there sufficient light elements to counterbalance the heavy? When tasting the rum, is there a succession of different stimulating aromas and flavours, or just one? Can you tell where the rum has come from, how it was made, its individuality and personality?

Pick up your glass and nose it gently. Remember that this is a spirit and there is likely to be a hot whack of alcohol giving you some nose burn. Note down the initial sensation: is it sweet or dry? Light or heavy? Is it hot or gentle?

Now nose again: is it fruity or floral? If it is fruity, what kind of fruit: tropical, dried, or crisp? If it's spicy, which spices? Can you detect vanilla or chocolate coming from cask? Relax and let the aroma come to you. Do not rush or inhale aggressively; you'll anaesthetize your nose. The rum is your friend.

Now taste. Sip and think of the texture. Is it hot or smooth, dry or sweet? Does it spread across the tongue or move quickly? Now repeat. This time concentrate on the flavours and where they appear. You're confirming what you have smelled but also sussing out balance. Each rum will have its own character, and the flavour should develop and change.

Now add some water and repeat the process. The water helps to reduce the alcohol and also trigger the release of aromatics.

Always be conscious of elements that work against the harmony of the rum. Is it really greasy? That's fusel oils. Is it neutral? That's vodka, not rum. Is the finish sweet, suggesting a heavy-handed addition of sugar?

Now relax – and most of all, enjoy. Rum, like any spirit, has been made to put a smile on your face.

HOW TO DRINK RUM

Why mix? Mixing becomes second nature when rum's involved, which is to its advantage. Unlike some other spirits, people know it works in mixed drinks. The question here is which mixer, and why?

To take the latter first, it's because rum offers up a range of flavours and characters. To simplify, each one will give fruits, flowers, and spices as well as sweetness and dryness – hopefully all in balance. Fruits give you mid-palate softness; flowers, top notes; spices provide interest on the finish. All of these will find their own way to interact with the mixer.

When a mixer is added, all of these facets should be enhanced. A successful mixer doesn't dilute or close flavours down, but opens them up, adding its own qualities to build on existing ones or create new flavours and textures. Carbonation adds pep and can lengthen; acidity will cut through thickness (and balance sugar). There's a complex and fascinating interaction taking place when you slosh something in a glass.

MIXERS

So, which ones to use? I asked a bunch of rum-loving friends. Thankfully, we were pretty much in agreement as to the four that follow. All the mixed drinks were made at a 2:1 mixer-to-rum ratio. They're rum drinks!

Coconut Water

Yes, it's good for you, but let's face it: you're not going to lose weight, balance your potassium levels, or become fitter by drinking rum and coconut water. You will drink an amazing combination, however. Coconut water is subtly sweet (check for sugar levels and go for stevia if possible) but also savoury and slightly salty/mineral-tinged, with a firm, nutty, element. In other words, there's complexity. The nuts link with oak; the sweetness links with fruit; the savoury notes add breadth, and many rums do have a mineral edge. It was the top performer. It has to be cold and it must be balanced.

Coconut water is a classic rum mixer.

Ginger beer adds pep and spice to a mixed rum drink.

Clementine Juice

Fruity juice was a no-brainer. Which one was a trickier proposition. Passion-fruit works, but it, and mango, can be a little thick. Pineapple was too... pineappley; grapefruit too sharp; and orange juice too sweet. Clementine juice hit the spot. The clementine, actually the smallest variety of tangerine, is a cross between the Mediterranean mandarin and sweet orange. This then places it within the orange flavour camp, but with a tartness that nods toward grapefruit. The acidity was the key, giving cut-through, opening up the rum, while the tropical nature of the fruit found a natural partner. It worked, but it was how it worked with aged rums that surprised the most.

Ginger Beer

Here we find a partner for spice and introduce carbonation. Ginger adds zing and ups spice, while pulling out those floral top notes and lengthening the finish. I'm a ginger beer fanatic, and for me Fever-Tree makes the best. It's a brewed blend of three types of ginger: fragrant from the Ivory Coast, intense Nigerian, and earthy Cochin. Cane sugar is the sweetener and high carbonation provides fine bubbles, allowing persistence.

Cola

Often the rum drinker's default drink. Cola has vanilla – there's a link; it has red fruit – another. It has sugary depth and softness. Did it work? Overall it was the lowest-scoring mixer, but when it worked it made excellent and quite grown-up drinks. The best rum 'n' coke was created by my buddy Ryan Chetiyawardana (see below).

RUM 'N' COKE FLOAT

50ml (1⅔fl oz) gold rum

20ml (¾fl oz) cola syrup (use the concentrate from a bag-in-box cola)

1 egg

Dry shake, then shake with ice, double-strain into a Coca-Cola contour glass, and garnish with a lime twist.

A Daiquiri offers a stern test for a white rum's quality.

RUM COCKTAILS

The rums then had to be given a run-out in a cocktail. Simple drinks work best in this context; they allow the spirit to shine, but as you'll see, they also offer up a sterner test than you might first imagine.

DAIQUIRI

This was chosen for (non-agricole) white rums because, well, what else would you do? The Daiquiri is a classic mix, but it is also a stern taskmaster, revealing as it does the balance, quality, and complexity of the rum. If you can't taste the base spirit and if it isn't also enhanced in the drink, then you move on.

Inevitably, for a simple drink there are myriad ways to make one. I wanted a ratio that would be balanced, with a sufficient kick of lime to energize (and reveal imbalances) and low enough on sugar to let the rum come through.

I wasn't going to borrow (and probably destroy) the wife's mixer to make these. Anyway, I have a penchant for the shaken Daiquiri; what's more, I have a 15-year-old daughter who needs to learn how to bartend. Don't call it child exploitation; call it learning a life skill.

60ml (2fl oz) rum
....
20ml (¾fl oz) lime
....
15ml (½fl oz) 2:1 simple syrup (*see* p.208)
....
cubed ice
....

Shake all the ingredients together over cubed ice and strain into glass.

CAIPIRINHA

Well, what else to use for cachaça?
Punchy yet smooth, vibrantly green
yet sweet. A great, great drink. This
is how I like 'em.

¾ of a lime, cut into wedges

15ml (½fl oz) 2:1 simple syrup
(see p.208)

cubed ice

60ml (2fl oz) cachaça

Gently muddle the lime and simple
sugar in an Old-Fashioned glass.
Add ice cubes and cachaça.

OLD-FASHIONED

This is so simple that on paper it might seem unlikely you
could learn any more about a rum by making one. Think again.
The sugar shows how sweet the rum already is; the bitters
can impose themselves, showing up bitter notes from over-
extraction and lack of balance. When it works, everything
combines seamlessly to reveal the rum in its most perfect state.

1 tsp 2:1 simple syrup (see p.208) or 1 sugar cube

3 ice cubes

6 drops The Bitter Truth Old Time Aromatic Bitters

6 drops Angostura Orange Bitters, or orange peel

60ml (2fl oz) rum

The old-fashioned way to make the Old-Fashioned
involves dripping the bitters onto the sugar cube, with a
splash of water, then muddling the rest of the ingredients
together. I had to make 69, so I took a shortcut of putting
the simple syrup and one cube of ice in the glass, dripping
the bitters in, stirring, then adding the rum, more ice, then
another stir, and a rest before drinking.

TI PUNCH

My first day in Martinique. I sit in the hotel lounge and a little
tray is brought out. On it a bottle of agricole blanc, *sugar, and*
some limes. "Ti Punch?" I was asked. "Oui, merci," Then they
walked away. I sat, wondering if they'd forgotten something
but, non: I had to make my own. That's how it's done and it's
a wise thing because everyone has a different tolerance for
alcohol, sweetness and, indeed, lime. This, then, is my ratio
for my kinda Ti Punch. Yours might be slightly different.

60ml (2fl oz) *rhum agricole*

10ml (⅓fl oz) cane syrup

lime peel, cut into an oval shape

cubed ice

Stir the ingredients together over ice in a Old-Fashioned
glass. Cut the lime peel with only a small amount of the flesh
so that the oils come into the drink rather than the juice.

RUM FLAVOUR MAP

The downside of rum's diversity is how tricky it is to navigate through the myriad different styles. I thought a rum flavour map might work. The aim here is to give an idea of that diversity, with each rum's position dictated by its aroma and taste.

HOW THE MAP WORKS

The vertical axis runs from "Fresh" at the foot to "Oak" at the top. Because the map is in two dimensions (but rum exists in three), it works in two ways. The first of these is wood influence.

As you move up the axis, so the influence of wood increases. Unsurprisingly, this means that white rums are congregated around the bottom reaches. Some white rums have, however, been aged in oak and then had the colour filtered out, meaning that not all white rums sit in a straight line.

At around three-quarters of the way up the lower quadrant, some additive flavours from wood begin to appear; this is where dry oak and then vanillin begin to emerge. Once over the horizontal line the oak steadily becomes the dominant player. By the top, the tannins from oak will have become increasingly prominent.

At the same time, the vertical line also measures an increase in weight of spirit, with the lightest rums at the very bottom, the richest toward the top. Distillate weight and general "richness" are therefore brought into account, as well as oak influence. This means that light, clean, column-still rums are at the very bottom of the line, while and heavy, pot-still rums are situated toward the upper reaches. It's also why some white rums are higher up the line than others – they have more distillate weight.

The horizontal axis also works in two ways. In simple terms, as you move from left to right, so the levels of sweetness in the rum increases. Rums that are distillate-

Rum is diverse – how can you navigate it?

driven are on the left-hand side of the vertical axis. This means that rums with dominant notes of fresh cane juice lie on the far left-hand side.

Not only is *rhum agricole blanc* found here, but so are its aged expressions, as cane juice notes are still predominant. Unaged and aged versions from the same producer are often in alignment vertically, demonstrating the existence of a distinct distillery character.

As you move toward the right, more fruit flavours begin to develop, reaching notes of tropical fruits and honey toward the centre.

Pungent pot-still rums, particularly those from Jamaica, are also on the left-hand side of the line, as their ester-driven notes are distillate-led and have an aromatic bridge to the characteristics of cane juice *rhums*. Although aged examples do show greater fruit and leathery weight, this reveals itself as a general enriching of the rum and doesn't shift them dramatically to the right.

After you cross the central line, the most dominant characteristic is sweetness. This could manifest itself as fruits, jam, or be sugar-derived. Sweeter white rums are therefore in the bottom right-hand quadrant. Equally, a rich, sweet, cask-aged rum – Demerara style or Latin – will sit on the far right of the upper right-hand quadrant.

OAKY & RICH

CRISP & DRY

LIGHT & FRESH

Admiral Rodney Extra Old

Rhum Rhum Liberation 2015

Appleton Estate Rare Blend 12-Year-Old

Plantation Rum Jamaica 2001

Mount Gay XO

Plantation XO 20th Anniversary

Santiago de Cuba Extra Añejo 25-Year-Old

Doorly's 12-Year-Old

Neisson 2004 Single Cask (bottled 2015)

Mount Gay Black Barrel

Rhum JM XO (bottled 2014)

Karukera Rhum Vieux Réserve Spéciale

Cockspur VSOR

Cruzan Single Barrel

Smith & Cross

Rhum JM 2003 (bottled 2014)

Havana Club Selección de Maestros

RL Seale's 10-Year-Old

Caroni 15-Year-Old, Velier

Penny Blue XO Single Estate Batch 004

Cuvée Homère Clément, Hors d'Age

Caroni 1999 (bottled 2015), Rum Nation

Banks 7 Golden Age Blend

Flor de Caña 12-Year-Old

St Lucia Distillers Chairman's Reser

Bally 2000

Appleton Estate Signature Blend

Bally Rhum Ambré

The Duppy Share

Ron de Jeremy

Compagnie des Indes Latino

Compagnie des Indes Jamaica Navy Strength 5-Year-Old

Havana Club 7 Años

The Real McCoy 5-Year-Old

Trois Rivières VSOP Réserve Spéciale

Bundaberg Small Batch

Don Q Gran Añejo

Barbancourt Réserve Spéciale ★★★★★ 8-Year-Old

St Nicholas Abbey 5-Year-Old

Elements Eight Gold

Barceló Imperial

Reserve Rum of Haiti, Distilled 2004, Bristol Classic Rum

Ron Montero Gran Reserva

Rum-Bar Gold 4-Year-Old

* Savanna Cuvée Spéciale 5-Year-Old

Bielle Rhum Vieux (Hors-d'Age) 4

Barceló Gran Añejo

Habitation Velier Foursquare 2013 (bottled 2015)

Abuelo Añejo

Mezan XO Jamaica

Blackwell Black Gold

OVD Old Vatted Demerara

Myers's Original Dark

Lamb's Navy Rum

Havana Club 3 Años

Rum Fire White Overproof

Caña Brava 3-Year-Old

Santa Teresa Claro

Rhum Rhum PMG

Wray & Nephew White Overproof Rum

Rum-Bar White Overproof

Rhum JM Blanc

Neisson Rhum Blanc

* Savanna Lontan Grand Arôme

Karukera Rhum Blanc

Brugal Carta Blan

Clément Première Canne

Bally Rhum Blanc

Clarke's Court Pure White Rum, Overproof

Elements Eight Platinum

Clément Canne Bleue 2013

Don Q Cristal

Clairin Sajous

El Dorado 15-Year-Old

Dos Maderas, Luxus Doble Crianza

Zacapa Solera Gran Reserva 23

Pampero Aniversario
Reserva Exclusiva

Dos Maderas PX 5+5

Hechicera   El Dorado 12-Year-Old

Diplomático Reserva
Extra Añejo

Bacardí Facundo Eximo 10 Años

Abuelo Añejo 12 Años,
Gran Reserva

Pusser's Gunpowder Proof Rum

Santa Teresa 1796 Antiguo de Solera

Rum Nation Peruano 8-Year-Old

Cacique 500 Extra Añejo
Gran Reserva

Cartavio XO

Brugal 1888 Gran Reserva Familiar

Cartavio Solera 12-Year-Old

**SOFT &
SWEET**

Bacardí Carta Ocho, 8 Años

Botran Solera 1893

Amrut Old Port

Matusalem Clásico   Old Monk 7-Year-Old

Gosling's Black Seal Bermuda Black Rum

Angostura 1919

McDowell's No 1
Celebration

Santiago de Cuba Ron Carta Blanca

Botran Reserva Blanca

Matusalem Platino

Bacardi Ron Superior,
Heritage Limited Edition

Tanduay White

KEY

White & Overproof Rums

Aged Latin-style Rums

Aged English-speaking
Caribbean Rums

Rhum Agricole, French
Départements, & Haiti

World Rum

Navy & Dark Rum

NOTE:
* For consistency's sake, the two *rhums*
from Savanna were included in the
French *Départément* tasting section of
the book. They are however both made
from molasses, hence their different
coding on the map.

SPICING AT HOME

You might be wondering where the spiced rums are. I tasted a selection and the quality was so poor that I took the executive decision to drop them. Anyway, it's more fun to make your own.

SPICED RUM

Empty a bottle of light gold rum into a Kilner jar and add the following:

1 vanilla pod

3 cloves

1 cinnamon stick

5 allspice berries

5 black peppercorns

1 star anise

¼ tsp grated nutmeg

4 pieces of ginger

the peel of 1 orange

Seal and store, giving it a turn every day. It should be ready in four to five days. Adjust for sugar levels to taste, then strain and bottle.

RHUM ARRANGÉ

Every bar and restaurant in the French départements has a vat of this in a prominent position filled with fruits or spices. Packets of spices are also available from a number of suppliers on the internet. Search for *"preparation pour rhum arrangé"*.

A French variant on the basic spiced recipe above is based on 2 litres (3½ pints) of white *rhum*, so double up on the ingredients, but add a chilli pepper, pinch of cumin, five loquat seeds, and 10 teaspoons of brown sugar and leave for three months.

Or just macerate one or two fruits, such as pineapple. In the past "pineapple rum" could either mean an ester-rich Jamaican rum that smelled of pineapple, or a rum that had pineapples macerated in it. It was big in the eighteenth and nineteenth centuries.

Pour a bottle of gold rum, and 50g (2oz) sugar into a large Kilner jar. Stir until the sugar has dissolved. Add a thumb-sized piece of ginger and a super-ripe pineapple cut into chunks. It's ready in a day – but you can leave for longer if desired. (Thanks to Ryan Chetiyawardana's *Good Things to Drink with Mr Lyan and Friends* for that tip).

Or… if you are really in a rush, buy a bottle of Plantation's Pineapple Rum (the "Stiggins Fancy 1824 Recipe" label). It really is magnificent.

THE RUMS

The more you look at this spirit, the more you realize that there is no such thing as "rum". Rather, there are seemingly infinite variations on the idea of rum.

Rums can be clear, golden, or black; they can be light or heavy, flavoured or spiced. They can be single cask, or a multi-nation blend. Rums can be drawn from gunk, or syrup, or juice. They can be imbued with the spirit of Dr Funkenstein, or be extra light. They can be drunk straight off the still, or aged. The wood used can be new, old, ex-Bourbon, ex-Sherry, ex-Cognac. Ageing can be static or solera, or a combination of both.

Rum has a fluidity unsurpassed in the world of spirits. No other spirit does that. That is rum's greatest asset. Here, then, are 110 of the world's best to explore.

WHITE & OVERPROOF

There's a belief that white rum is like vodka: that things only get serious when the spirit is put in cask. By extension, this means that if a recipe calls for a white rum you can grab the first bottle that comes to hand and it will do. After all they're interchangeable, aren't they? Well, no. Sure, there are some that teeter on the edge of neutrality, but the majority have their own individuality, character, and – like people – likes and dislikes.

Molasses or cane syrup, multiple column or pot still, different yeasts, cask-ageing, and filtration, and tradition: all of these combine to give a world of rums which are, by and large, lighter but no less lacking in character; it's just often at a more subtle level. I say "often" because here also lurk overproof rums and you wouldn't dare to walk up to one of them on a dark night and call it a wimp.

Overall, clementine juice and coconut water proved the most amenable partners for these rums; the former ruled in overproof world, while cola, so often the default mixer for white rum, was the least inspiring.

Daiquiris proved to be a drink that gives rum a proper test. While the simple mixes could enhance, meld with, or sometimes cover the rum, there is no hiding place with a Daiquiri; the drink succeeds or fails thanks to the rum.

BACARDÍ RON SUPERIOR, HERITAGE LIMITED EDITION

44.5%/37.5% ABV

Bacardí uses its own strain of yeast to induce a rapid fermentation. Distillation is in a five-column system. Two distillates – a light *redistilado* (95% ABV) and a heavier, fruity *aguardiente* (80%) are produced, filtered, then aged separately, in ex-Bourbon casks, for at least one year. The rum is then charcoal-filtered to remove colour.

I compared the higher-strength Heritage bottling with the standard strength. The nose of the Heritage is light, floral, with a white mushroom note and a sooty element. The palate is rounded and fruity, slightly sweet, with a crisp backbone and a slightly peppery end. The standard strength is lighter, sharper, and vanilla-accented.

Coconut water links well with the floral notes of the Heritage and adds nuttiness, while the higher alcohol stops the clementine juice becoming oppressive. Ginger beer brings out a mineral note. This Bacardí makes a damn fine Daiquiri; the lower strength lacks impact. Again, the floral elements come through and the alcohol adds texture, not burn. Everything is toned down when the standard strength is used, yet it performed decently across the board. The only cavil would be the weird bubblegum effect given by cola. A go-to rum, and once you've tried the higher strength, you'll never go back.

TASTING VALUES			
4/2.5	Coconut Water	4.5/3	Clementine Juice
5/3.5	Ginger Beer	3/3	Cola
4/2.5	Daiquiri		

BOTRAN RESERVA BLANCA
40% ABV

Guatemala's Botran is made from sugar-cane "honey" (for which read "syrup") and uses a yeast isolated from pineapple. Fermentation is long – up to 120 hours – which helps develop more complex esters. Distillation is in copper-rich columns and ageing takes place in a solera before filtering.

A dry, biscuity opening quickly sweetens alongside a medicinal note (crepe bandage, plaster of Paris) before things become more creamy, with banana skin and lemon tones. It seems to shift between dry and sweet: now crisp, now perfumed, like potpourri. The palate is sweet and slightly fat in the centre, while the back palate shows some drier and fresher elements: even a little cherry.

It comes into its own when mixed. Coconut water gives this an almost vinous, green, slightly sappy quality with a hint of smoke. Cola's weight is kept in check by that dry quality, while ginger beer is fair. The pick of the bunch is the clementine juice, where the purity of the juice helps to pull out the fruits. Keep it short, though.

The Daiquiri picks up the lemon and runs, slightly knocking the nose askew, but the palate is settled, with the rum adding softness and a perky little gasp of surprise on the end. In other words, a good match, with some complexity.

TASTING VALUES			
4.5	Coconut Water	4.5	Clementine Juice
3	Ginger Beer	4.5	Cola
4.5	Daiquiri		

BRUGAL CARTA BLANCA
40% ABV

Brugal is so all-pervasive in the Dominican Republic that you wonder at times whether it's used instead of holy water to baptize babies. It starts with molasses (with a minimum of five per cent fermentable sugars), and a two-day fermentation, followed by double distillation in stainless-steel columns, giving a base spirit of 95% ABV.

The white is light to medium-bodied, with hints of biscuit before it perks up into lime zest, green mango, and a little waxiness. The palate is clean, with a soft and lightly sweet mid-palate and balancing dryness, which develops into pear and a hint of coffee. The weight shows with water and it is this heaviness that makes coconut water a little ponderous as a mix.

Cola, however, is an ideal loosener for a night's activities, while ginger beer adds an almost peppery element to what starts smoothly. The acidity of the clementine juice gives a good cut-through and takes you to the beach and a game of dominoes.

When thrown into a Daiquiri, that palate weight helps to add some tropical-fruit-accented layers. A cool, clean, drink with just enough sweetness to balance. I'd happily have a couple.

TASTING VALUES			
2.5	Coconut Water	4.5	Clementine Juice
3.5	Ginger Beer	3.5	Cola
4	Daiquiri		

CAÑA BRAVA 3-YEAR-OLD
43% ABV

Made in Panama in the Las Cabres Distillery under the watchful gaze of Francisco "Don Pancho" Fernandez, who left Cuba to settle in Panama in the 1990s. This uses local molasses, fermented with Don Pancho's own yeast and distilled in a five-column set-up, being collected at between 92% and 94% ABV. Ageing starts in new American oak barrels for 18 to 24 months, then in used barrels for a further 12 to 24 months.

Fresh and fragrantly complex, the nose shows grassy notes with a fruity undertow of cooked pears and lemon blossom, with understated but evident cask notes. The palate is medium-weight, allowing some light raspberry-leaf elements and a fat, rounded, balanced mid-palate with touches of chocolate.

This is one of those rums that breezily welcomes all comers; even the cola is pleasant, if unremarkable. Coconut is enhanced by it, imparting a breadth but also calmness and control, and it's that element which also benefits ginger beer, allowing the spice to become part of the overall mix. It creates a mouthwatering mix with clementine juice. The rum slips with ease into a Daiquiri: dry and clean aromatically, but with the right degree of richness to add weight to the palate, while the touch of sweetness balances the citrus. Classy in all departments. A benchmark.

TASTING VALUES				
4.5	Coconut Water	5*	Clementine Juice
4.5	Ginger Beer	3	Cola
5*	Daiquiri		

DON Q CRISTAL 40% ABV

This molasses-based Puerto Rican rum uses the firm's own yeast, which was isolated in the 1930s. A 48-hour ferment yields a wash, which is distilled in a five-column still. The rum is then aged for between 18 months and five years before being filtered.

It's crisp, whistle-clean and almost chalkily dry, with hints of light green fruit. The palate is quite thick, with a sweet centre, some citrus, and then a fresh, long, slightly spicy finish. In other words, a well-made, classic, modern, Puerto Rican style that begs to be mixed.

The coconut water slightly dominated, but the mix was fresh and clean, unlike the cola, which stomped with gay abandon the rum's delicate nature. It was the clementine juice and the ginger beer, however, that allowed this rum's racy charms to show themselves.

Although light in character it worked well in a Daiquiri, the rum adding balance, a fresh top note, and giving just enough drive. I'd head in the direction of the shaker every time.

TASTING VALUES			
3	Coconut Water	3.5	Clementine Juice
3.5	Ginger Beer	2	Cola
4	Daiquiri		

ELEMENTS EIGHT PLATINUM
40% ABV

Produced to its owner's specification in St Lucia, Elements Eight was one of the first white rums to try and break into the premium market. It is a blend of three different stills: column, traditional pot, and hybrid pot/column Vendôme still, made with three different yeasts and producing 10 different bases. These are then aged for four years before blending and filtration.

It's light and slightly edgy when neat; tight, with mineral elements mixed with whittled sticks and a steely edge before the molasses oozes forward. Water allows a herbal lemon note to develop. The palate is broader, with some tightness on the sides, but a more subtle, fruited, baked-apple element shows itself. The finish is long – quite a change from the initially lean nose.

A good performer overall, though with cola it's a bit boring and flat; the clementine juice picks up the pace. Ginger beer goes a step further with a zesty drive, but richness in the mid-palate, making it more than the sum of its parts. Head, though, for the coconut water; the green notes in this mixer find a natural partner and you can imagine the gentle sigh of the two as they meld together.

Things are different with the Daiquiri. If Katharine Hepburn were a drink, it would be this. The palate has drive, energy, and that steeliness. It depends on how you like your Daiquiris – or your actresses, come to think of it.

TASTING VALUES				
5	Coconut Water	3.5	Clementine Juice
4	Ginger Beer	3	Cola
4	Daiquiri		

HAVANA CLUB 3 AÑOS 40% ABV

Produced by the complex Havana Club blending regime, this uses *aguardiente* and high-strength cane alcohol. There is a single base (see p.57) for 3 años that is blended with the cane alcohol, which itself has been aged for a minimum of three years. This is pale straw with a dry, crisp, carnation-like nose with just a whiff of molasses, lemon, and plum skin. It's pretty heady, mixing oils, almond, and frangipane with a mineral, almost saline note, then green leafiness. The palate is clean, dry, and lightly acidic, adding to that intense citrus experience. Oak injects a gentle support – clean and balanced – with a dry, lightly spiced, controlled mid-palate. A clever rum.

A great all-rounder, this transports you to Havana's Malecón, at different times and moods. With ginger beer it's a surprisingly subtle and complex drink; it's the oak I suspect. Cola provides a fairly grown-up mix and is best with some lime, while the coconut water picks up the earthy underpinnings, enhancing the blend. The opposite happens with the clementine juice, where acidity is key in bringing the juice into the rum – again, a relatively serious drink.

The weightier elements of the rum help give a Daiquiri weight, power, and balance. This is a must-have.

TASTING VALUES			
4.5	Coconut Water	4.5	Clementine Juice
5	Ginger Beer	4	Cola
5*	Daiquiri		

MATUSALEM PLATINO 40% ABV

Although there is Cuban ancestry here, Matusalem is bottled in the Dominican Republic. The nose is also quite different to the dry Cuban style. Imagine gorging yourself in a French patisserie and you're on the right track here. There's cream, vanilla extract, butterscotch, lychee, then (strangely) tobacco. It's thick and sweet on the tongue, with masses of butterscotch, some sultana, and a buttery, slightly oily finish. Treat it as a liqueur white rum.

It's often drunk with cola, but I'd advise against that, unless you want a drink which smells like the interior of a new car – and, hey: some people might be turned on by that. Ginger beer goes all herbal, with the fatness of the rum suppressing the spicy bite and making things smell like a pick 'n' mix counter. The clementine juice's acidity helps give some balance, but the top performer is coconut water, which calms the rum down, adds exotic top notes, and gives you a proper drink – albeit one like a melting coconut ice cream.

Here is a Daiquiri disaster though: the result is all candyfloss and strawberry.

TASTING VALUES			
4	Coconut Water	3.5	Clementine Juice
2.5	Ginger Beer	2	Cola
1	Daiquiri		

THE RUMS: WHITE & OVERPROOF

SANTIAGO DE CUBA
RON CARTA BLANCA 38% ABV

This enigmatic Cuban rum is slowly beginning to be exported – which is no bad thing on this showing. All Cuban rums have to be aged for two years, which gives this a slight lemon cast and an aroma that mixes apricot blossom with a waxy perfume, which brings to mind scented candles. It's fat and slightly oily on the tongue, where a candied element comes through, followed by rose water, lime jelly, and Turkish Delight notes alongside some sugary sweetness. There's a strange vision of an old-fashioned sweet shop by the roadside.

Ginger beer falters, throwing out whiffs of swimming pools. Coconut water is a better, more balanced mixer, with some mineral tones, albeit adding some oiliness on the end. Cola makes a classic Havana (or Santiago) Cuba Libre with some weight and a hint of oil. The top performer here is clementine juice, which is clean, slightly bittersweet, enhanced, and balanced.

It makes a fair Daiquiri, with that heavier distillate adding notes of bamboo and lime flower. The palate has some richness, but overall it's short, so keep this small, hard, and cold.

TASTING VALUES				
3	Coconut Water	4	Clementine Juice
2	Ginger Beer	3.5	Cola
3	Daiquiri		

SANTA TERESA CLARO 43% ABV

This molasses-based Venezuelan rum has been aged for a minimum of two years, and by the look of its straw colour, only given a light filtration at the end of that period. It also means it's not exactly *claro*. A waxy note starts things off, alongside some sweetness that brings to mind single cream and custard tarts. The oak gives a very gentle grip and maybe just a touch of char, and there is a firmness to the mid-palate, where things start to become sweetly spiced, with hints of cumin.

If cola gives a slightly bizarre aroma of braking on a hot road, the other mixers perform admirably. The slight extra weight given by oak-ageing is key here. There's even a hint of molasses with the juice, while ginger beer moves things into the realms of fresh Indian spices – and greater length. Coconut water is the best match, providing an extra layer to a moderately complex rum, with some coconut flesh coming into play and just sufficient sweetness to make this an excellent pairing.

The cask adds light vanilla and woody elements as a Daiquiri, making for a slightly confused nose, but the palate has verve and weight. It might be a little short but this is an all-round performer you can rely on.

TASTING VALUES			
5	Coconut Water	4.5	Clementine Juice
4.5	Ginger Beer	3	Cola
4	Daiquiri		

TANDUAY WHITE 36% ABV

There are rums that become part of a country's culture – think of Wray & Nephew Overproof or Bundaberg – but there's only one that has a 99 per cent share of its home country's rum market. That's Tanduay, which sells in excess of 16 million cases a year in the Philippines alone. This, though, represents a fall from its height of 21 million cases, and is one reason why you are beginning to see it on the export market.

The firm was started in 1854 by Don Jose Joaquin Ynchausti, Joaquin Elizalde, and Juan Bautista, who began to make rum at an existing distillery located in Hagonoy, in the province of Bulacan. In 1869, a new distillery was built at Isla de Tanduay, close to Manila. The Elizalde family remained in charge until 1988, when the company was purchased by Dr. Lucio C. Tan, who initiated a massive expansion of the distillery.

The white is clean, very light, and clearly well distilled. Fresh to start with, there are touches of citrus, then vanilla ice cream, with a slightly ginny, pine element in the background. It's sharper than other light rums. The palate is like drinking alcoholic American cream soda while munching on Dolly Mixtures. It doesn't like water; it's a low-strength bottling.

Perhaps unsurprisingly, all of the mixers dominate. They make pleasant enough drinks – the coconut water shows the most rum element – but they're not rum drinks. The same applies to the Daiquiri.

TASTING VALUES			
2.5	Coconut Water	2	Clementine Juice
2.5	Ginger Beer	2	Cola
1	Daiquiri		

CLARKE'S COURT PURE WHITE RUM, OVERPROOF 69% ABV

The Grenada Distillery has been making rum in Woodlands Valley since 1937, and this is its main product: molasses based, high in strength, clear as water. You get intense notes of fresh green beans, drying citrus peels, artichoke, and molasses – a pure distillate aroma which, with water, brings to mind smelling angelica in a hot tin shed. The palate has heat, but there's a natural sweetness: orange, kumquat, tropical fruit. Warm, weighty, and with water-soft raspberry, it dries well. It has style, power, balance, and length.

The drinks here were, wisely, given greater dilution, but there's no need to approach this with any trepidation. Clarke's Court proves to be an amenable and charitable host. Cola is ripe and sweet, and though coconut water starts edgily, the distillate drives the mix, making things clean and subtly dry. Ginger beer has a natural affinity, picking up those top notes and adding lime and spice, while the clementine juice matches the fruity elements, as the rum cleaves a passage through its thickness, giving an energetic, fresh finish.

Things become trickier with Daiquiris; the strength now makes it harder to balance, with lime proving a problem. It's not a drink for the faint-hearted; Hemingway would have loved one of these. Acknowledging that thought, I went for grapefruit juice instead and lo: it worked.

TASTING VALUES			
4	Coconut Water	4.5	Clementine Juice
4	Ginger Beer	3	Cola
3	Daiquiri		

RUM-BAR WHITE OVERPROOF
63% ABV

The Worthy Park Estate is located in Lluidas Vale, in the heart of Jamaica, and has been producing sugar since 1720. The Clarke family is only the third to own it, and its members have been in charge since 1918. It is one of the few rum producers to be self-sufficient in molasses. The distillery, like many in Jamaica, was forced to close due to low demand, but distillation restarted in 2005. It adheres to traditional methods, using pot stills to make a range from light to high ester. Rum-Bar is a blend of three of these, fermented with three yeasts: proprietary, wild, and distiller's. No dunder or muck pits are used.

The esters come forward immediately, with whiffs of gloss paint, pineapple, and banana chews backed with light oils. It's heady and complex, with everything turned up to 11. On the palate you get light sultana, fresh coffee bean, and a pleasant dusty earthiness. Anise comes out with water, alongside liquorice and molasses. This is really well made.

Avoid cola, though. Ginger beer has a sight antiseptic hit initially, but it works, as does coconut water, which allies itself with the sultana to make a rich drink. The clementine juice walks in, calms the situation, and adds balance, giving a palate-cleansing quality that enhances the fruit and soothes the alcohol. The funk, which isn't apparent in the other mixes, springs forward in a Daiquiri. Keep it simple is the key here.

TASTING VALUES			
3.5	Coconut Water	5	Clementine Juice
3	Ginger Beer	2	Cola
N/A	Daiquiri		

RUM FIRE WHITE OVERPROOF
63% ABV

Jamaica's Hampden Estate, located in Trelawny Parish, is one of the world's most important rum distilleries and the repository of traditional rum-making techniques: dunder, muck pits, long wild yeast ferments that can rumble on for up to three weeks before being distilled in elephantine pot stills, whose trunks plunge into two retorts. It is funk central, hogo heaven, and was kept alive for many years, thanks to E&A Scheer of Amsterdam and its belief in traditional rum character for blends. The Hussey family has owned Hampden since 2009, stabilized the business, invested in production, and started to bottle the rum rather than only selling in bulk. The signature style is Rum Fire White Overproof.

The nose moves along the tropical-fruit axis, with lots of pineapple and a rich, controlled funkiness that gives a layering effect. Behind that is a seductive sense of decay, alongside fresh baguette and pear notes. Pineapple syrup leads off the palate, alongside a hint of nail polish, then a long, biscuity sweet element. Water smooths things, adding liquorice, sweet cicely, and star anise.

Cola has a whiff of shoe polish; coconut water sulkily sets itself up in opposition; ginger beer is as edgy as a tense dominoes match; but the clementine juice rides in throwing orange, pineapple, and mango to the crowds. Use it for floats, in tiki drinks, with juice – but not in a Daiquiri, where the pineapple elements go feral.

TASTING VALUES			
2	Coconut Water	4.5	Clementine Juice
2.5	Ginger Beer	2	Cola
N/A	Daiquiri		

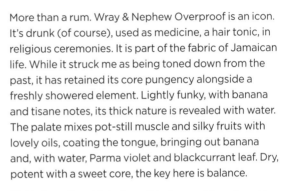

WRAY & NEPHEW WHITE OVERPROOF RUM 63% ABV

More than a rum. Wray & Nephew Overproof is an icon. It's drunk (of course), used as medicine, a hair tonic, in religious ceremonies. It is part of the fabric of Jamaican life. While it struck me as being toned down from the past, it has retained its core pungency alongside a freshly showered element. Lightly funky, with banana and tisane notes, its thick nature is revealed with water. The palate mixes pot-still muscle and silky fruits with lovely oils, coating the tongue, bringing out banana and, with water, Parma violet and blackcurrant leaf. Dry, potent with a sweet core, the key here is balance.

It's that quality which allows it to be a solid performer when mixed. Banana leaves are added to coconut water; there's a menthol hit with ginger beer, fresh and fine; slight phenolic funkiness with cola, but also the start of some dark cherry which works in a guilty-pleasure fashion. It's overproof and clementine juice that, again, prove to be the killer: all juicy, freshly picked basket of fruits, energetic, and – most importantly – balanced. You could drink this all day. I wouldn't recommend it for health-and-safety reasons, but you could.

Just as you thought the wilder elements of the rum were under control, along comes the Daiquiri. "The bananas have got drunk and had a party," said my daughter. She's right. Try it sometime.

TASTING VALUES			
3.5	Coconut Water	5	Clementine Juice
4	Ginger Beer	3.5	Cola
3	Daiquiri		

AGED LATIN-STYLE

The first question here is: can you find some degree of commonality between rums produced with molasses, sugar-cane syrup, or cane juice, which are made in column stills, or pots and columns, and which could then be aged either statically or in solera? Where's the connection, Dave, other than the fact that the main language of each of these countries is Spanish?

That's the world of rum. You come across similar issues when trying to corral all the rums of the English-speaking Caribbean. Latin-style rums can be considered as a style – albeit a wide-ranging style. If the English-speaking Caribbean's roots are in the pot-still rums of Jamaica and Barbados, then their Latin colleagues take their cue from the Cuban rums that appeared in the nineteenth century: lighter in character, column-still based. It's a definition which still holds true.

In general terms, the younger-age-profile rums (up to Gran Añejo or equivalent) were more amenable that their older colleagues.

The most consistent mixer here was coconut water; the others have their own highs and lows. There's no surprise about this, as coconut water's dry edges help to add firmness when put with rums that are often, but not automatically, sweet (some teeth-achingly so). On the other hand, it was interesting to see that a number were dry.

In other words, don't generalize, and treat each rum on its own merits: is it balanced, is it complex? There are some phenomenal examples here that are essential for any rum-lover.

ABUELO AÑEJO 40% ABV

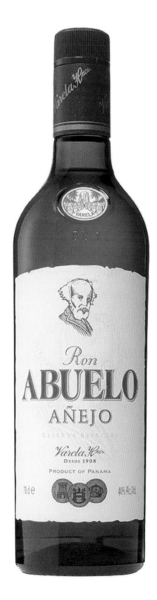

Varela Hermanos started in 1908, when José Varela Blanco started processing sugar at the Ingenio y Destileria San Isidro (San Isidro Plant and Distillery) in the cane fields around the town of Pesé, in central Panama. Rum began to be distilled in 1936. Today the firm, now run by the third generation, also produces gin and liqueurs.

The nose is clean, with spearmint coolness and freeze-dried raspberry soon followed by powdery spices. It's a young rum, and this freshness provides a bracing cleanliness, with the wood never dominating but rather giving delicate support and aromas of pencil shavings. It has a medium-sweet start, and though initially lighter and sweeter than the nose suggests, it's balanced, with a full back palate.

Clementine juice is pleasant but not, well, rummy enough, and ginger beer is spicy and subtly weighted. Coconut water and cola both work well, which is unusual, but they do so in different ways, showing that there's a complex rum at the heart of things. The cola kicks out vanilla notes, giving balance as well as a nutty link to the dry elements in the rum, meaning the mix never becomes too cloying. Coconut also works, with the oak adding grip and restraint, but not preventing this from becoming a rich, unctuous mix. You do, however, need more oak-driven complexity and weight to make a great Old-Fashioned.

TASTING VALUES			
5	Coconut Water	3.5	Clementine Juice
3.5	Ginger Beer	3.5	Cola
3.5	Old-Fashioned		

BACARDÍ CARTA OCHO, 8 AÑOS 40% ABV

You know something is happening in a market when one of its biggest players moves into what had been a "specialized" area. Products like Bacardí Ocho show that momentum is building in the aged-rum sector. While the casks used are all ex-Bourbon barrels, different numbers of fills are employed to age the two Bacardí distillate's, giving a reserve stock from which to draw for bottlings such as this, where the youngest element will have been in cask for eight years. The nose has a mix of light honey, sandalwood, and a herbal hint that leans into caraway. It's thicker in the mouth, quite fat and spiced, with Seville orange and apricot tones. Water reveals a mix of custard and cinnamon buns in the centre, before toffee pudding finishes things off. Sweet certainly, but balanced.

There's something about ginger beer that struggles with many Latin-style rums. Here, the ginger rockets away from the rum. Clementine juice is controlled and has balance, even elegance; while cola starts off sweet, the palate has richness and much-needed grip. Coconut water just sidles in and makes a delicious drink, extending the rum's flavours while adding a silky, nutty texture. In an Old-Fashioned, there's citrus and a real pick-up on the middle of the palate, where crisp oak, pulpy fruit, and a new coriander-leaf note appear. This is sweet, so be careful when balancing.

TASTING VALUES			
5	Coconut Water	4.5	Clementine Juice
3	Ginger Beer	4	Cola
3.5	Old-Fashioned		

BOTRAN SOLERA 1893 40% ABV

Botran uses sugar-cane honey (syrup) for its rums and ages in solera, and you wonder whether that base ingredient helps to contribute to the rum's sweet and, yes, quite syrupy opening. There are touches of crème brûlée, kumquat, tayberry, fresh tobacco, and clean, crisp oak that then lift, leaving a haunting note of a beeswax candle. The palate shows the very Guatemalan mix of sweet banana, undercut with a biscuity crispness to start with before sweetness develops. Water brings out some light cumin to start, and again, things get sweet by the middle.

When I started to mix, I wondered if someone had put tomatoes in my coconut water, alongside a sprinkling of desiccated coconut. Cola, meanwhile, is jammy and short, then ginger beer heads off into the sweaty jungle and the clementine juice struggles to find a path between distillate, fruit, tannin, and spice.

The Old-Fashioned, upon whose shoulders a lot was riding, was fresh, but there's sugar in the way as well as some figgy notes which show promise, but then it splits: sugar goes one way and oak the other, with no hope of reconciliation. One for the ice cube, methinks.

TASTING VALUES			
3	Coconut Water	3	Clementine Juice
2.5	Ginger Beer	2.5	Cola
2.5	Old-Fashioned		

THE RUMS: AGED LATIN-STYLE

CARTAVIO SOLERA
12-YEAR-OLD 40% ABV

Cartavio master blender Frederico Schulz blends together 85 per cent column-still and 20 per cent pot-still rums for this bottling, which is then aged in a solera, comprising a mix of American, French, and Slovenian oak casks. The "12 years" is slightly confusing as soleras operate on the principle of the cask never being emptied, so take it as an estimated calculation of the time it takes for new rum to pass through the system.

As with the Gold, the nose is quite dry, with hints of oak leaves, Brazil nuts, and *horchata*, also a dried-fruit sweetness that mingles with acacia honey, garden bonfires, and sandalwood. The palate shows Sherry-like date notes before it gets sweeter. It remains a little blunt when neat, so add water to taste its full range.

Cola is pretty much *the* mixer for most drinkers, but if you value your teeth, steer clear. Opt instead for the lightly sweet but still fresh mix achieved with ginger beer, the marzipan-like fruity depths of clementine juice, or – best of all – the mix of gentlemen's club and, oddly, Kola Kubes with coconut water, which has a certain degree of gravitas.

Things fall apart with the Old-Fashioned, which is too sweet to be balanced. Sip with ice is the best option.

TASTING VALUES			
4.5	Coconut Water	4	Clementine Juice
3	Ginger Beer	2.5	Cola
2.5	Old-Fashioned		

THE RUMS: AGED LATIN-STYLE

COMPAGNIE DES INDES
LATINO 40% ABV

Compagnie des Indes was established by Florent Beuchet, with the intention of paying homage to the rums that were imported by the various European trading companies of the eighteenth and early nineteenth centuries. All the rums – which can be single casks or blends – are bottled in France. There's no caramel colouring, but sugar levels are declared if used.

Latino is a blend of 60 per cent Guatemalan rum and 40 per cent from Guyana, Barbados, and Trinidad. *Latino* therefore refers to a style rather than a place of origin. It's also a blend of molasses, cane syrup, and cane-juice distillates, mostly from column still (there's a little Bajan pot in the mix), aged, statically in mostly new American oak. Pale in colour, it opens with clean, round, but biscuity Guatemalan notes, with subtle, fleshy fruits behind, given a top note by menthol, then *crème pâtissière* and light oils. The palate is candied and sweet.

Cola is too sweet, collapsing the mix; the rest, however, shine. Coconut water brings out an interesting cigar-like note, adds length, and enhances the rum's freshness. This is what helps give the clementine an uncharacteristic focus, linking well with the tropical-fruit notes. Ginger beer has a big, head-clearing retronasal blast, but shows good depth and a brightness. The Old-Fashioned, on the other hand, starts well but ends up too sweet.

TASTING VALUES			
4	Coconut Water	4	Clementine Juice
5	Ginger Beer	2.5	Cola
2.5	Old-Fashioned		

THE RUMS: AGED LATIN-STYLE

CRUZAN SINGLE BARREL
40% ABV

St Croix's Diamond Rum Company was founded by Malcolm Skeoch, who restarted a distillery on a sugar plantation he had bought in 1910. Although a period of Prohibition-enforced silence soon followed, the Skeoches fired up the stills again in 1934, with Cruzan as the main brand. By the 1950s, it was an ultra-modern distillery whose pots had been replaced by columns, themselves further refined by Herminio Brau, who had run Puerto Rico's Rum Pilot Plant's labs. The Nelthropp family has managed the firm since the 1960s.

The single barrel has a big, Bourbonesque opening: think vanilla, banana smoothie, pecan syrup. It's thick, bold, and coconut-accented with jags of lemon candy and orange zest. The palate is sweet with caramelized fruit and, toward the finish, spice and caramelized red fruits. It's rum done Bourbon-style.

The issue here is whether the oak, which is mighty, acts as a barrier or a contributor to a greater whole. By and large it works well. Clementine juice struggles a little on the palate; cola gets an extra buttery edge and a chunky mid-palate; ginger beer shows an unusual muscularity linking with the spicy end. The coconut water takes the wood notes, absorbs them, then spins them into new formulations, making a more complex rum drink. You won't go far wrong here if you like your Bourbon Old-Fashioneds on the sweet side.

TASTING VALUES			
4.5	Coconut Water	3.5	Clementine Juice
4.5	Ginger Beer	4	Cola
3	Old-Fashioned		

THE RUMS: AGED LATIN-STYLE

HAVANA CLUB 7 AÑOS 40% ABV

It's fair to say that this was one of the new rum world's game-changers. What had been a category that was either Naval style or white was now also proving it could make complex sipping spirits. This example also underlined the inherent quality of classic Cuban rum. Made from a blend of "bases" that are themselves a blend of different-strength distillates, the youngest element here is seven years old; the oldest, twice that.

You get immediate nuttiness from oak, then warm coffee notes, light criollo chocolate, an earthy and dry depth, suggesting this is good sipping material. In time, there's molasses. The broad, naturally deep palate shows touches of orange peel, cherry, black grape, roasting spices and finally, on the drying finish, that chocolate note once more.

This was one of the few across-the-board winners. Coconut water makes things almost Fino Sherry-like, enhancing that Cuban minerality. It's clean, soft, with a lovely harmonious finish. Clementine juice picks up the citrus, then adds to mid-palate weight; ginger beer is relaxed and laid back, allowing a slow unfolding of sweetness, then allying itself with the spice. Cola – making a Cubata – brings the chocolate and dark fruits forward, but the dryness of the rum gives balance.

An Old-Fashioned provides a platform for all the elements seen individually with the mixers to work together, creating a complex drink. An essential rum.

TASTING VALUES			
4	Coconut Water	4	Clementine Juice
5	Ginger Beer	5	Cola
5*	Old-Fashioned		

LA HECHICERA 40% ABV

The brainchild of Laura and Miguel Riascos, La Hechicera ("The Enchantress") is blended and matured in Barranquilla, Colómbia, by *maestro ronero* Giraldo Mituoka Kagana. It is a blend of different (unspecified) Caribbean rums that have been aged in an American oak cask solera.

From the start, it's clear that there's mature stock being used: leather, spice, char, a touch of vanilla and chocolate are here, along with some treacle. It's complex, and while oak-driven, it's not woody. There's a funky depth and layered quality to its broad, deep palate that opens with a pleasantly bittersweet tone, though it's slightly blunted by some sweet, green-fig jam when neat. Water helps reveal its complexity and structure – and also cuts back the sweetness.

It works across the board with mixers, except with cola, which amplifies the leather but is also oddly short. The rum brings weight to ginger beer, which changes into stem ginger, and though it needs just a touch more length it's a minor quibble. Clementine juice works well with the softening oak and pulling out the core, but it is coconut water that most amply shows this rum's complexities: sweet fruits, coconut flesh, char, dryness, and persistence. The Old-Fashioned is complex, with the bitters seamlessly adding a root-like, exotic element. Very finely balanced, with a light grip before a spicy/citrus ending. A must-have.

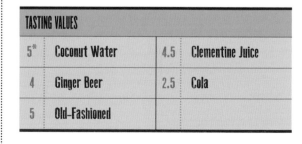

TASTING VALUES			
5*	Coconut Water	4.5	Clementine Juice
4	Ginger Beer	2.5	Cola
5	Old-Fashioned		

THE RUMS: AGED LATIN-STYLE

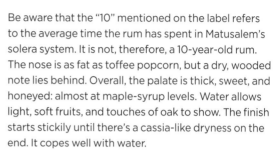

MATUSALEM CLÁSICO 40% ABV

Be aware that the "10" mentioned on the label refers to the average time the rum has spent in Matusalem's solera system. It is not, therefore, a 10-year-old rum. The nose is as fat as toffee popcorn, but a dry, wooded note lies behind. Overall, the palate is thick, sweet, and honeyed: almost at maple-syrup levels. Water allows light, soft fruits, and touches of oak to show. The finish starts stickily until there's a cassia-like dryness on the end. It copes well with water.

Coconut water shows flesh notes as well as the maple syrup, making this a pretty, calorie-rich mix. Cola pulls out the vanilla and slathers it back on the tongue in a very thick fashion. Ginger beer is, unsurprisingly bigger and sweeter than normal but has a remarkable effect of making things so gingery you can't help sneezing. More control is found with the clementine juice where acidity and sweetness achieve some balance.

This is a light and slightly perfumed rum, which causes some conflict with the bitters in an Old-Fashioned, so go lightly.

TASTING VALUES			
3	Coconut Water	4	Clementine Juice
3.5	Ginger Beer	3	Cola
3	Old-Fashioned		

RUM NATION PERUANO 8-YEAR-OLD 42% ABV

Fabio and Walter Rossi started as independent Scotch whisky bottlers (Wilson & Morgan) in the early 1990s, but like all of their Scottish equivalents had a keen interest in rum. In 1999, Rum Nation started. The firm, based in Treviso, bottles by region/country of origin rather than distillery. Peruano is a column-distilled, molasses-based Peruvian rum sourced from Lambayeque.

It's a thick middleweight with aromas of crème de cacao (my immediate thought was to make a Mulata) over blackberry and an aroma akin to warm massage oil. The palate starts with clove, then dry tannins, liquorice, and masses of molasses. Water shows drier elements coming through.

It proves to be a decent performer, the coconut water giving some richness, though it's too sweet for my palate and becomes slightly cloying. Clementine juice also ends up being too chunky because of the sugar, but ginger's zip helps to cut through. Cola, too, goes well, working alongside the darker fruits. It's a big and bold mix.

As an Old-Fashioned, the rum, now slightly cherry-accented, only comes through on the tongue before an interesting note of hickory, smoke, and the baize of a deserted pool hall emerges. Sweet.

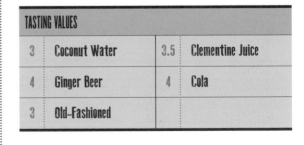

TASTING VALUES			
3	Coconut Water	3.5	Clementine Juice
4	Ginger Beer	4	Cola
3	Old-Fashioned		

THE RUMS: AGED LATIN-STYLE

SANTA TERESA 1796
ANTIGUO DE SOLERA 40% ABV

A highly regarded solera-aged rum, this seems to have become slightly sweeter over the years. There's certainly a glossy aspect to the nose, where there are slightly sweet stewing fruits and a note akin to Trappist beer, then comes stewing citrus. It seems weighty with a little crunchiness, hard toffee, and, with water, a massive tangerine hit along with turmeric. The palate is thick, deepening into raisin and a Sherry-like oxidized nuttiness. Water makes it quite light and citric.

This is a well-constructed rum with some elegance. The ginger in the beer is tamed and amalgamated into the mix; the clementine juice in its turn adds a freshness and clarity, while cola has a fuller mid-palate that equally works nicely. It's when we reach coconut water that things become that little bit more serious, with the oak (which can be overlooked) acting as a bridge to the off-dry nature of the mixer.

The Old-Fashioned is massively citric, which is no bad thing as long as it can be controlled. There are spices from oak and the bitters working in tandem, and the pulse of the rum carries the middle – but then it gets too sweet. I'd keep it with ice, to be honest.

TASTING VALUES				
4	Coconut Water	3	Clementine Juice
3	Ginger Beer	3.5	Cola
3	Old-Fashioned		

ABUELO AÑEJO 12 AÑOS, GRAN RESERVA 40% ABV

Given static ageing in ex-Bourbon barrels, this opens with lush, ripe, dark-fruited notes akin to stewed prunes and what we might as well call Panamanian funk, which comes across like chamois leather, then hints at old attic dust and, with evidence of long ageing, the cigar box. The palate follows pretty much the same journey, with rum-soaked dried fruits, Christmas cake mix, and surprisingly little grip for such a prolonged period in cask. Water demonstrates its density, and there's this pleasant bittersweet note of black-cherry jam at the end.

The rum, as you've seen, is mature and dense, and it is one of those that just prefers to be left on its own – that's certainly the case with the simple mixers. Ginger beer ends up tasting like green tea, but it does help reveal the tannic structure. Coconut water shares this slight astringency, clementine juice dominates, and when cola is brought in, the rum drives forward but the cola kicks back like a sulky toddler.

In a sense of resignation you make an Old-Fashioned, only to discover that the rum's true complexity emerges. It's now drier and the whole package has the air of slightly faded grandeur as that funk returns, adding to this sense of elegant decay. A real find.

TASTING VALUES			
2	Coconut Water	3	Clementine Juice
2	Ginger Beer	2.5	Cola
5	Old-Fashioned		

BACARDÍ FACUNDO EXIMO 10 AÑOS 40% ABV

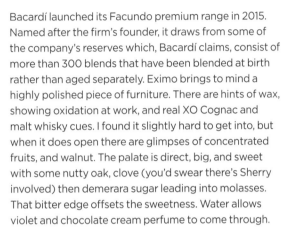

Bacardí launched its Facundo premium range in 2015. Named after the firm's founder, it draws from some of the company's reserves which, Bacardí claims, consist of more than 300 blends that have been blended at birth rather than aged separately. Eximo brings to mind a highly polished piece of furniture. There are hints of wax, showing oxidation at work, and real XO Cognac and malt whisky cues. I found it slightly hard to get into, but when it does open there are glimpses of concentrated fruits, and walnut. The palate is direct, big, and sweet with some nutty oak, clove (you'd swear there's Sherry involved) then demerara sugar leading into molasses. That bitter edge offsets the sweetness. Water allows violet and chocolate cream perfume to come through.

As you take this step up in terms of age profile, the rums move away from general geniality and into a realm where less interference is the best option. Cola here is all perfumed rose essence; clementine juice can't shift the oak; ginger beer somehow resembles elderberries and, while effervescent, is astringent. Only coconut water comes out with any credit, folding the rum into itself, showing good integration, length, and some oak. The Old-Fashioned, on the other hand, sees the oak pushing forward with the additions adding to its supple power and amplifying a chocolate tone. One for Scotch-lovers.

TASTING VALUES			
4.5	Coconut Water	3	Clementine Juice
3.5	Ginger Beer	2	Cola
5	Old-Fashioned		

THE RUMS: AGED LATIN-STYLE

BARCELÓ GRAN AÑEJO
37.5% ABV

This family firm was founded in the Dominican Republic in 1929 by Julián Barceló, although it wouldn't be until 1950 before the first rums were released. The third generation of the family still sits on the board.

This is a cane juice-based, column-still rum distilled to 95% ABV in a four-column set-up, then aged statically. It is also bottled at low strength, which, as we'll find out, leads to some issues. The nose is all florist shop at closing time, with praline hints and a natural, dry, oaky note. The palate is light and clean, which, while a little dusty at the start, sweetens in the middle where the citrus flies out along with banana, apricot, and more of that clean oak.

This low strength does it no favours when it's mixed. Cola is minty, somehow; ginger beer picks up a strangely fishy note, and the mixer seems to be providing all of the energy to drive it across the palate. This passivity is repeated with coconut water, which brings out some banana skin and a nuttiness, but everything falls down in the middle, as does the clementine juice, which is the best of the bunch.

As an Old-Fashioned, things are light, clean, and citric, with a little sugar. What a difference a few degrees make.

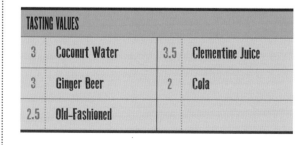

TASTING VALUES				
3	Coconut Water	3.5	Clementine Juice
3	Ginger Beer	2	Cola
2.5	Old-Fashioned		

BARCELÓ IMPERIAL 38% ABV

Created in 1980 as the top rum at the time in the Barceló range, this cane-juice-based rum shows an appealingly plump nose with some candied stone fruits, light leathery hints suggestive of age, then toasted marshmallow and ginger. The sweeter elements are there but not dominant. On the tongue it becomes a little slick, with notes of potpourri, citrus, lavender, and chocolate, before stem ginger creeps through on the end. It remains slightly ethereal, thanks to the lower strength. It doesn't like water, so what happens when you have to add it, even if it is with mixers? Sadly, it dies on you.

In the Old-Fashioned I like the wormwood note and the dry, bosky, leafy quality, but it becomes really sweet and unbalanced.

Low alcohol might save a distiller money, but alcohol carries flavour, and cutting to under 40% ABV loses texture and aromatics. I'd have this with one very hard ice ball.

TASTING VALUES			
2	Coconut Water	N/A	Clementine Juice
N/A	Ginger Beer	N/A	Cola
2.5	Old-Fashioned		

BRUGAL 1888 GRAN RESERVA FAMILIAR 40% ABV

The Edrington Group's purchase of Brugal in 2008 didn't only result in a Scotch distiller learning about another category, it also gave the rum firm access to years of expertise in oak, especially in the realm of Sherry casks (Edrington owns The Macallan and Highland Park). The first evidence of this was the 1888, which is given up to eight years in medium-toasted, ex-Bourbon casks. The rum is then placed into first-fill Spanish oak ex-Oloroso casks. The nose opens to coconut, raisin, then country-house-style aromas, but with the vibrancy needed for a rum like this. There's an oxidized Sherry edge, which takes you into a damp-floored headily aromatic *bodega* – Sherry funk! The palate is sweet, nutty, and intense, with layers of baked stone fruit, and treacle. It's rich, but not sweet. There's a difference.

I thought the sherried elements would be too much for the mixers. On the contrary: the coconut water brings an extra, mellow layer with low tannin allowing things to flow over the tongue elegantly. Clementine juice has life and energy, but ginger beer falters when the oak emerges. Cola, which on paper should have worked, promises a lot on first sip, but the rum soon gets bored and leaves the party by the side exit.

The Old-Fashioned is deep, with masses of tempered chocolate. This unhurried mix deserves respect.

TASTING VALUES			
5*	Coconut Water	4	Clementine Juice
3.5	Ginger Beer	3	Cola
4.5	Old-Fashioned		

CACIQUE 500 EXTRA AÑEJO GRAN RESERVA 40% ABV

Made by Destilerias Unidas in Venezuela (where the Cacique brand is the top seller) this Gran Reserva shows slight abrasiveness on the nose, with green, lemony elements. It becomes considerably sweeter with a drop of water. The palate is thick and highly concentrated, with masses of damson jam, some figgy elements, black pepper, raisin, and a heavy sugar impact on the end. Water brings out a touch of candyfloss on the finish.

This is a pleasant, if simple and sweet rum, and its lightness works best in tandem with coconut water, which seems to be made for it as the green notes act as a link. It's way better than the cola, which pulls out a farmyard note before falling into a puddle of sugar. There are rounded elements with a slight – and needed – kick from the clementine juice, while ginger beer adds grassy elements and just a little touch of bitterness at the end, but not unpleasantly so.

The Old-Fashioned quickly heads into caramel toffee land, with cinnamon accents. It's fine, but has split into constituent parts by the end.

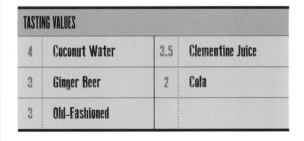

TASTING VALUES			
4	Coconut Water	3.5	Clementine Juice
3	Ginger Beer	2	Cola
3	Old-Fashioned		

THE RUMS: AGED LATIN-STYLE

CARTAVIO XO 40% ABV

As we move inexorably up the Cartavio range, so the aromas become more concentrated and the wood just that little bit more apparent. The nose here mixes praline with acacia-honey-dipped almonds, then mothballs and old wardrobes. The palate is almost mentholated and very sweet, with stewed damsons, bitter orange, an old leather tobacco pouch, maraschino, and a sweet, treacly finish that tickles the throat with some charred oak. If you like your rum thick and slick, then you'll love it.

Cola relishes rums with this quality, but things can often get too sweet. This, though, teases out some oak and there are bittersweet edges, red fruits, and some length as well. Coconut water follows a similarly tricky path, allowing the oak frame to show itself, and while the nose of both the clementine juice and ginger beer mixes are excellent – summer gardens in the latter's case – there's little substance on the palate.

In the Old-Fashioned, that dry oak leads you on to a hot, dusty trail out of a spaghetti western before the sugar heaves itself out from behind a rock and guns our heroic rum down.

TASTING VALUES			
4	Coconut Water	3	Clementine Juice
3	Ginger Beer	4	Cola
2.5	Old-Fashioned		

DIPLOMÁTICO RESERVA EXTRA AÑEJO 40% ABV

You want sweet? I'll give you sweet, in a thick-caramel-sauce-over-rum-and-raisin-ice-cream-with-a-glass-of-cream-Sherry-on-the-side-kind-of-sweet, which some people absolutely love. The start of the palate, however, is much drier than you might expect, with some heavy pot-still distillate adding considerable heft and impact, and allowing some good layering of flavours. This is excellent distilling at work. Then the sweetening kicks in, moving things into tinned blackcurrant and cherry pie filling and fruit syrups.

This doesn't bode well for mixers, and ginger beer falls at the first hurdle, with musky elements showing. Clementine juice, though, shows a boisterous kind of charm, and while it might lack in coherence it has power. Cola offers more sugar than at a kid's birthday party, but coconut water imposes some sanity on proceedings, with simply lovely rich toffee and caramel, but cut with some nuttiness, the more exuberant elements being held in check as the mixer gets to work.

The Old-Fashioned mix needs those sugar levels controlled. The caramel notes are increased but citrus perks things up, while the bitters add a necessary edge before bouncy chocolate comes through. The sweet finish makes it less complex, but it's not bad at all.

TASTING VALUES			
4.5	Coconut Water	3.5	Clementine Juice
2	Ginger Beer	3	Cola
4	Old-Fashioned		

DON Q GRAN AÑEJO 40% ABV

Two different rum styles of up to 12 years old are used here. Light rums made with a short ferment and distilled in a five-column set-up are blended with heavy rums fermented for between one and two weeks, then put through a single copper column. A mix of American oak and ex-Sherry casks is used, as is static ageing.

Light in colour, the nose is delicately scented, with coriander seed, ginger, and lemon balm accents. The oak is precise and more restrained than most Gran Añejos. The palate is creamy, clotted cream at that, suggesting some first-fill casks, then come prickling spices and a hint of soft richness in the middle. Water shows a more polished note, followed by sage and lemon thyme.

Forget about trying it with cola, though there are some richer elements here that suggest a rum Manhattan might work. Ginger beer is enlivening, but not any better than the rum on its own, which is much the same with clementine juice. It's coconut water that has this extra dimension. Here, the crispness of the rum is brought out, then layered on the palate, which has a silkier texture but is still aromatically pure.

It might be an old-fashioned kind of rum, but it's not an Old-Fashioned one. There's lemon again here, with some cask smoke – but it's light. Sometimes it's best to leave things alone. Add ice and enjoy.

TASTING VALUES			
5	Coconut Water	3	Clementine Juice
3	Ginger Beer	2.5	Cola
3.5	Old-Fashioned		

THE RUMS: AGED LATIN-STYLE

FLOR DE CAÑA 12-YEAR-OLD
40% ABV

Five generations of the Pellas family have made rum at the San Antonio Sugar Mill, in Chichigalpa, Nicaragua. The original distillery was built in 1890, and still uses molasses extracted from the firm's own cane fields. There are no additives such as caramel tinting used in the rum, and ageing is static rather than via solera.

The nose is dry and quite green/herbal – think linden blossom, fruit-bush leaf – with cigar wrapper tones, a crispness, and pleasant pollen notes. The palate is full with a light toffee-apple element and the herbal edge of the nose obscured initially before there is a burst of kumquat, bergamot, that tobacco note and light oak.

This calm restraint carries through the mixes. If the cola is supercharged, it is also slightly fragmentary. Clementine juice is sharp and clean and finds a partner with the citrus, while ginger beer hits a lovely balance; if anything, the ginger is calmed down, offering just a little frisson of spice. Coconut water here also is kept within a delicately precise, clean, and balanced frame.

All of this is carried over into the Old-Fashioned, where the bitters add this slightly Gothic element, bringing out aromas of a Seville orange grove, but the clean elegance of the rum remains. Class in a glass.

TASTING VALUES			
5	Coconut Water	3.5	Clementine Juice
4.5	Ginger Beer	3	Cola
5	Old-Fashioned		

HAVANA CLUB SELECCIÓN DE MAESTROS 45% ABV

Originally called Cuban Barrel Proof, this rum undergoes a three-stage blending process. The bases are selected, blended together, and recasked for a set period in more active wood. After this period of maturation these casks are re-examined and the final blend is made. There's no dilution at bottling.

The nose is heady, citric, and floral, with fresh fruits dipped in honey. The oak adds light cedar touches and an aroma akin to Bakewell tarts (marzipan). There's still the mineral note, intense citrus notes, then fruit sugars that briefly suggest Cognac. A rum where spirit, air, and oak all work together equally. The palate is tight to begin, but later offers dried banana chips, dried peel, and sumac.

This is a bring-it-on rum that isn't fazed in the least by the occasional curveball a mixer can hurl. Clementine juice is the least impressive, a little woody but still fine; ginger beer is soft and clean with a zesty zap to the finish, while, for once, cola has good length and brings out a savoury element. It's coconut water where the greatest complexity can be seen: fusing citrus, fruit, and oak into a contemplative drink.

The Old-Fashioned kicks off with citrus before melding into some floral notes and a punchy palate that soothes down into a sophisticated afternoon drink.

TASTING VALUES			
5*	Coconut Water	3.5	Clementine Juice
4	Ginger Beer	4	Cola
5	Old-Fashioned		

PAMPERO ANIVERSARIO RESERVA EXCLUSIVA 40% ABV

Founded in Venezuela in 1938 by Alejandro Hernandez, Aniversario was launched in 1963 to commemorate Pampero's 25th anniversary. Snuggled in its little leather pouch is a little bomb of a bottle containing a rum which, on the nose, shows immediate funky maturity. There's semi-dried black fruits to the fore, along with overripe banana, molasses, date, raisin, and cake mix, which slowly moves into a resinous waxiness suggestive of age. It occupies a similar aromatic ground to Armagnac. The palate is solid and powerful to start, with deep, almost sooty/peppery Shiraz-like weight alongside balancing sweetness. The tannins are supple, and the deep earthiness of the nose moves to almost gamey notes, then Nutella. Bold and complex.

The first three mixers didn't work. This is a rum that you sit and sip, which refuses to stoop to such base behaviour – and quite rightly. Then came cola, which ends up creating a great, complex, mix of PX Sherry, black cherry, leather, and earth. You see, sometimes it's worth persevering.

This was borne out by the Old-Fashioned, which took all the crepuscular complexities of the neat rum and amplified them. It's a drink that begs for a cigar and demands to be experienced.

TASTING VALUES			
N/A	Coconut Water	N/A	Clementine Juice
N/A	Ginger Beer	5	Cola
5*	Old-Fashioned		

THE RUMS: AGED LATIN-STYLE

SANTIAGO DE CUBA EXTRA AÑEJO 25-YEAR-OLD 40% ABV

A quarter of a century is a hell of a long time for any spirit to spend in cask. Doing it in the heat of the Caribbean usually ends up with the drinker picking splinters out of their tongue. Here, though, there are still fresh, lifted aromas of honey, macadamia, mace, and cinnamon gently backed with a slight touch of the antique shop. It slowly begins to shift toward the Grand Marnier arena with just a little confected touch of old-fashioned cough sweets. The palate starts with cinnamon balls, cardamom, and clove oil. If you like spices and sweetness, you'll love this. Water makes things dusty and peppery.

It might seem heretical to mix a rum as rare as this, but rules are rules and it acquits itself well. Coconut water becomes perfumed with cigar wrappers; there's a similar aromatic story with ginger beer, which sings out in the mid-palate but fades dramatically – something shared with cola, which brings in liquorice and cherry but also slumps. Clementine juice, however, gives glints of what might happen if this were used in some extravagantly priced tiki drink.

You are on surer ground with the Old-Fashioned, which, while it still brings to mind the interior of an old house in Havana, has this lovely menthol note and, while delicate, has poise. The palate leads on orange and anise. Strangely compelling.

TASTING VALUES			
3.5	Coconut Water	4.5	Clementine Juice
3.5	Ginger Beer	3	Cola
4	Old-Fashioned		

ZACAPA SOLERA
GRAN RESERVA 23 40% ABV

The insanely complex maturation system invented by Lorena Vasquez for Zacapa has been outlined elsewhere. What we are concerned with here is what the rum aged above the clouds tastes like.

For starters, it's not as sweet as many think. Instead, it is layered and slightly smoky, with roasted coconut, vanilla pod, grapefruit, dry-roasted spices (kalonji, coriander), then raisin and clean oak and some of the beeswax note spotted in Botran. Then comes Assam tea, and autumn leaves. The palate is multilayered, with dried and tropical fruit notes but with some grip, cherry chocolate, cooked fruits, mulberry, and supple tannins. Balanced and complex, it finishes with an almond crunch before thick PX Sherry comes forward.

Impossible to mix with? Think again. Here ginger beer, which struggles with many Latin-style rums, blooms as flavour bridges form between citrus, sweetness, oak, and spice. This has verve and a long, complex finish. The rum takes the cola to college, but is a little severe with it when it arrives there. Coconut water comes across like eating a coconut in a jungle, while the clementine juice also is pulled deep. There's sweetness, acidity, and a mix that's enhanced. The Old-Fashioned expands into mint chocolate with just a little bit of extra sweetness coming through at the end before the PX unfolds itself. Singular.

TASTING VALUES			
4	Coconut Water	4	Clementine Juice
5*	Ginger Beer	4	Cola
4.5	Old-Fashioned		

AGED ENGLISH-SPEAKING CARIBBEAN

Here is a selection of rums that are mostly a mix of pot still and column – occasionally 100 per cent pot – made in a drier style than those in the Latin camp. Look for the deep, molasses-driven, sometime leathery richness on the nose and mid-palate, which shows pots at work. Because of no to low sugar addition you also get more obvious grip. The oak is therefore a major contributor to aroma, flavour, and structure which means the quality of the oak and the skill of the blender are paramount. Youth isn't being boosted by sugar; neither is over-wooded character being masked.

That is not to say that there is an identikit "English-speaking" style. There are clear differences in approach between each island/country. Jamaica proudly pushes its pungency. Barbados is softer, more fruit-driven and slightly citric, Guyana bridges the pot/column principles of the English-speaking Caribbean with Latin levels of sweetness. Each distillery has its own spin on that style.

Cola was the least-amenable mixer here, while clementine juice edged past coconut water. Jamaican rums worked better with ginger beer than those from Barbados, and the older the rum, the trickier the mixing became. Most of the rums in this group are best on their own or in cocktails.

Whatever your preference, though, be sure to enjoy and explore.

ANGOSTURA 1919 40% ABV

When a fire ripped through the Trinidad and Tobago Government Rum Bond in 1932, the few remaining casks – presumably now fairly heavily charred on both sides of the staves – were bought by JB Fernandes, master blender of Fernandes Distillers. All had been filled in 1919, and so a new rum was born. Today's 1919, now made by Angostura, is slightly different. It starts vanilla heavy, with crème caramel and toffee, white chocolate, a little peach, and a light oily note. There's more of the same of the palate, with some heavy, honeyed sweetness, then a sudden dip into dry spices, strawberries dipped into melted Oreos, and Ben & Jerry's Caramel Chew Chew.

The scores show that simple mixes work best here, but you need to be aware of the sweetness of the rum. Coconut water is great if you like Bounty bars; the clementine juice gives a certain control, but again, is sweetened. Ginger beer is slightly better because of its spice, but even this has to work hard. Cola is a good mix, again if you like sweetness and the two elements pulling in the same direction with extra vanilla.

The Old-Fashioned mix is like an explosion in an old-fashioned sweet shop.

TASTING VALUES			
3	Coconut Water	3.5	Clementine Juice
3.5	Ginger Beer	3.5	Cola
N/A	Old-Fashioned		

APPLETON ESTATE
SIGNATURE BLEND 40% ABV

The Appleton Estate claims to be the oldest sugar estate and distillery in Jamaica in continuous production. Lying in the middle of the island's Cockpit Country, its cane fields provide all of the molasses used for its range. Previously known as V/X, this is a blend of 15 aged rums (pot still and column) aged for an average of four years. The nose is unmistakably Jamaican, with tones of shoeshine stand, old banana skin, light leather, and tobacco, but there is also passion fruit and mango. Water adds elegance, slightly more spice, and a touch of oak. The palate is medium-weight and manages to balance the rich oils of pot-still rums with hibiscus and gentle, yielding fruits within a discreet structure.

This is a genuine all-rounder. Cola is solid, with its depths adding an extra dimension rather than just making things sweeter. The rum has the greater say with coconut water, lifting and fusing the mixer's sweet-sour green notes and playing all sorts of tunes in the middle of the mouth. Clementine juice prospers in drier rums, allowing the fruits to have a major say, while ginger beer is a delicious mix, where the spices come through, the ginger extending the finish and the rum giving softness to the mid-palate.

The Old-Fashioned shows more pot-still elements, but for once the rum seems light. Keep it to the simple mixes.

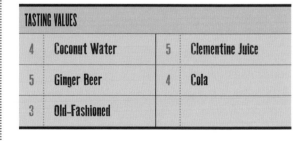

TASTING VALUES			
4	Coconut Water	5	Clementine Juice
5	Ginger Beer	4	Cola
3	Old-Fashioned		

THE RUMS: AGED ENGLISH-SPEAKING CARIBBEAN

COCKSPUR VSOR 43% ABV

The Cockspur brand was founded in 1884 by Valdemar Hanschell. Under Barbadian law, distillers and bottlers had to be separate entities, so Hanschell had to source liquid from across the island. Since 1973, Cockspur has been made exclusively at the West Indies Rum Distillery, where the island's first column still was installed. An old, two-column John Dore still is used there, and provides some of the rum for the blend. The rest is a mix of high-strength light rum from a four-column set-up, and medium-weight and heavy pot-still rums.

The nose is strangely reminiscent of the stewing dark fruits typical of Latin-style rums. There's fudge, and chestnut honey mixing a banana lift with dried papaya and milk sweets. More oaky, mature notes come through with water. The palate is quite dry, then sweetness like caramel toffee, with a light almond touch. This is a rich, wood-accented, commercial style.

It comes into its own when mixed, apart from with cola, which foams, then instantly flattens. Stick to the breadth given by ginger beer, which has just enough oak and spice to link, or the almost Bourbon-accented mix with coconut water that shows how well balanced the oak component is. This oak sits back a little with the clementine juice, which becomes absorbed into the mix.

In an Old-Fashioned, this rum shows how the drink prospers when there is structure to play with.

TASTING VALUES			
5	Coconut Water	5	Clementine Juice
4	Ginger Beer	3	Cola
4	Old-Fashioned		

COMPAGNIE DES INDES JAMAICA NAVY STRENGTH 5-YEAR-OLD 57% ABV

Here is Florent Beuchet's take on a high-strength Jamaican rum of the type drunk in days of yore. It's a blend of three pot-still rums and swaggers out with mid-weight funky richness that opens into primed canvas, dried pineapple, milk chocolate, and Brasso tones with a complex, smoky background note. The palate starts sweetly, with concentrated soft fruits, light oak, a touch of molasses, and a grapefruit-like acidity. The back palate then deepens into coffee and stewed peach. Tannins are supple. Overall, a medium-weight rum with elegance and a refreshingly uncluttered quality.

Mixers show how versatile Jamaican rums can be. Coconut water shows an almost ugly beauty: funky, nutty, long, and intense, and while not for everyone, I like it. Ginger beer brings out cinnamon and star anise, and is complex and long. Cola works because it becomes less of a mixer and more of an integrated element, which helps reveal the rum's darker side. Clementine juice adds another layer of complexity: a perfect base for tiki drinks. As an Old-Fashioned it's balanced, but the edginess of young(ish) pot stills comes through when bitters are added. You have to like the funk.

TASTING VALUES			
4	Coconut Water	5*	Clementine Juice
4.5	Ginger Beer	5	Cola
3.5	Old-Fashioned		

THE DUPPY SHARE 40% ABV

Created by George Frost and Jess Swinfen, Duppy (it's Jamaican for "ghost") is a blend of molasses-based three-year-old Jamaican pot-still and five-year-old Bajan column-still rums all aged in ex-Bourbon barrels.

Jamaica's to the fore to start with: all overripe fruity depth, new suede jacket, oolong tea. After this initial blast, things soften and out come pomegranate and guava and a gentle calm descends. On the palate, it's very gentle and soft to start, with the pot still taking you into the tropical undergrowth, where berry fruits and wet leather shine, while sweeter tropical elements, muscovado sugar, and (when watered) floral notes fly above. Restrained and balanced.

If cola does little to excite, the rest pick out different facets of the rum's dual personality. Clementine juice attaches itself to the fruitier elements, but also adds some acidity, and gives a mellow result; ginger beer does the opposite and pulls out the funk. With coconut water, everything returns to a state of balance, mixing elegance and heft, sweetness, a dry note, richness and aromatics. Simply a gorgeous drink.

Old-Fashioned-wise, there's enough mid-palate weight to add some power, giving a mix that's nutty with banana peel, citrus pith, lychee, and violet touches. Great rum.

TASTING VALUES			
5*	Coconut Water	4	Clementine Juice
4	Ginger Beer	3.5	Cola
4	Old-Fashioned		

ELEMENTS EIGHT GOLD 40% ABV

Made in St Lucia, this gold rum – and I like the fact that it proudly calls itself that – is a blend of 10 different six-year-old rums that have been distilled in column, pot, and hybrid pot-and-column stills, then aged in ex-Bourbon barrels.

As with the brand's white rum, there's a tightness to the unreduced nose. This then loosens into green olive tones, before apple syrup comes forward. It's got energy, but seems relatively light. It's more of a palate rum, to be honest, opening slowly and showing more ripe, fleshy, tropical-fruited weight along with mandarin and molasses, then cut flowers. All in all, clean and stylish.

The mixers were slightly, well, mixed. Coconut water is rich and slightly sappy (that green-olive note) with some power, if lacking in length. Cola is straight-ahead, while clementine juice melds nicely with the fruits, bringing out an almost Sémillon-like vinous quality. This is quite a laid-back and slightly restrained rum, and it benefits from the assertiveness provided by the ginger beer.

It's this cool quality that makes the Old-Fashioned seem a little austere to start with, but persevere with it and you'll get a pretty sophisticated drink with little explosions of flavour from the bitters and a longer, spicier, if ultra-relaxed, finish. It's good.

TASTING VALUES			
3.5	Coconut Water	3.5	Clementine Juice
4	Ginger Beer	3	Cola
4	Old-Fashioned		

HABITATION VELIER FOURSQUARE 2013 (BOTTLED 2015) 64% ABV

This forms part of a range launched by the irrepressible Luca Gargano of Velier, which aims to bring transparency into rum. The label describes the still type (pot with double retort), cask type (ex-Cognac), distillation, and bottling dates, the angel's share (15 per cent), and the fact it's sugar-free with no caramel or chill-filtering.

A two-year-old pot still is a serious rum-drinker's tipple, and this rum delivers. The nose is vibrant but complex, with that pot-still brassy note, drum-skin tones, and molasses richness backed with Bajan purity and freshness. With water there's turmeric root and asafoetida, bamboo, and plantain. The palate is soft and easy to drink, with a tongue-clinging quality. Fine-boned and dry when neat, it needs water to bring out deeper notes of preserved lemon and a finish of phenolic pot still and molasses. The cask influence is low: some spiciness.

It's too punchy for coconut water; the pot still makes cola go off-dry; and ginger beer is an imperfect fit. Clementine juice, however, settles things down nicely. There's still a dry edge but the heat has gone. In an Old-Fashioned, there's a strangely resinous, curried element and a firm edge. Use in complex tiki drinks, or sip on its own.

TASTING VALUES			
2	Coconut Water	5	Clementine Juice
3.5	Ginger Beer	3.5	Cola
3	Old-Fashioned		

MOUNT GAY BLACK BARREL
43% ABV

A blend – like all of the Mount Gay range – of column- and pot-still rums, Black Barrel has a higher percentage of pot in the mix. After blending, the rum is given a second period of maturation in heavily charred barrels: hence the name.

The nose shows mature, quite rich pot-still aromas, with some honey and light molasses, then at the back there is some dry spice and a hint of oak. It's all very intriguing and slightly vinous, with ripe – even overripe fruits – and a touch of tamarind, baked banana, and light tropical fruit. The palate shows more oak presence, a mix of crème brûlée and spice as well as light coffee. Water lightens it, but also allows tobacco, citrus, and more fruit to emerge.

Here, cola actually makes for a quite delicate and even subtle mix, with good mid-palate plumpness. Ginger beer brings out mixed toffee and fresh ginger – again, with this mellow mid-palate. Clementine juice is easy-going, while coconut water edged things, thanks to a slight firmness that then shifted to citrus, a fuller palate, and hints of sweetness on the finish.

An Old-Fashioned is clean and orange-accented with the wood toned down. It needs a little more weight for this serve, but is still a fine performer.

TASTING VALUES			
4	Coconut Water	3.5	Clementine Juice
3.5	Ginger Beer	4	Cola
4	Old-Fashioned		

THE RUMS: AGED ENGLISH-SPEAKING CARIBBEAN

THE REAL MCCOY
5-YEAR-OLD 40% ABV

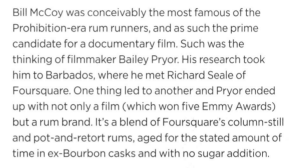

Bill McCoy was conceivably the most famous of the Prohibition-era rum runners, and as such the prime candidate for a documentary film. Such was the thinking of filmmaker Bailey Pryor. His research took him to Barbados, where he met Richard Seale of Foursquare. One thing led to another and Pryor ended up with not only a film (which won five Emmy Awards) but a rum brand. It's a blend of Foursquare's column-still and pot-and-retort rums, aged for the stated amount of time in ex-Bourbon casks and with no sugar addition.

Bajan balance is key here, with some deeper pot-still notes mingling with crème anglaise, fried banana, lightly caramelized fruits, and a drier calamus back note. As it opens there's more pot still, and cashew mixed with dried tropical fruits. The medium-bodied palate has a fair hit of oak lactone, soft fruits, and nutmeg on the end.

It's a rum that's made for mixing and does so without losing its character. You get garam masala with cola; a perky lift with the clementine juice, which imbues it with richness. Stem ginger and extra length come across with the ginger beer, while coconut water has that degree of sophistication that comes from having the dry nature of oak to link with. An Old-Fashioned has more depth, but I found it tricky to balance.

TASTING VALUES			
4.5	Coconut Water	4	Clementine Juice
4	Ginger Beer	3.5	Cola
3	Old-Fashioned		

RON DE JEREMY 40% ABV

As this rum is named after an adult-movie performer (so they tell me; I'm not familiar with his oeuvre) you might expect it to be little more than a marketing idea where all the effort went into the (clever) name and less concern was given to the liquid. That's not the case. This is a fine rum. It used to be a Panamanian product, but is now a blend of pot and column still and rums from Barbados, Trinidad, Jamaica, and Guyana. The oldest component is eight years old.

The nose is clean, with some caramel and fudge packed in, before a fair amount of good wood comes through followed by a spurt of tropical milkshake. In time, there's a touch of herbal freshness. It has a surprisingly gentle touch. It fills the mouth with custard and a firm nuttiness, then dives into banana-split territory, providing a good silky flow with just enough firm oak to stop it flopping about.

Coconut water brings out sweet coconut flesh and caramel, and is a little too fat. Clementine juice shows anise bite and some mango, which, if short, delivers. Ginger beer is spicy and long, with just a hint of a confected note, while cola lingers well – and if the overall effect is of a cola float then there is some harmony when they come together.

The Old-Fashioned manages to be both sweet and have whiffs of cigar smoke. The vanilla then pumps through to a slightly messy climax.

TASTING VALUES			
3.5	Coconut Water	4	Clementine Juice
3	Ginger Beer	3	Cola
2	Old-Fashioned		

RUM-BAR GOLD 4-YEAR-OLD
40% ABV

A blend of different types of pot-still rums from Jamaica's Worthy Park Estate (*see* p.83). Rum-Bar Gold is aged for a minimum of four years in ex-Tennessee whiskey barrels. This age is regarded as being the optimum point for the estate: where the distillate and oak are in ideal balance. It's borne out here. There's the estate's clean, but pot-still-rich, character; some light, dry oak; even some delicate ash wood-type elements and a hint of deeper funkiness. The palate starts off with wild fennel, which then goes into fennel pollen and boiled sweets, Starbursts, and a deep syrup note cut with almond in the middle. The finish is clean, dry, and not confected.

Worthy Park's elegant take on Jamaican funk prospers here. Clementine juice is clean, and if short, remains a good drink. Coconut water is green to start with (think privet hedge), then hazelnut and floral notes appear. Ginger beer adds length, while cola, so often the trickiest of mixers, here rolls over and offers up raisin, Griottine cherry, and mulberry.

An Old-Fashioned is light and clean, with an earthy edge and decent balance between dried fruit and banana, and there's that fennel and marzipan edge again. It may lack weight to carry, but it's not bad.

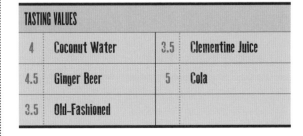

TASTING VALUES			
4	Coconut Water	3.5	Clementine Juice
4.5	Ginger Beer	5	Cola
3.5	Old-Fashioned		

THE RUMS: AGED ENGLISH-SPEAKING CARIBBEAN

SAINT LUCIA DISTILLERS CHAIRMAN'S RESERVE 40% ABV

One of the most innovative rum producers, Saint Lucia Distillers maximizes the potential of its distillery to produce eight different marks of rum. Two types of raw materials (molasses and cane juice), two yeast strains, and three still types are used. Its Coffey still makes three marks: one for white, two for ageing. One of its John Dore pot stills makes rum from cane juice and molasses, while the other just uses molasses. The company's Vendôme pot still uses both cane juice and molasses.

The Chairman's Reserve is a blend of pot- and Coffey-still rums with an average age of five years, given between nine and 12 months in new oak before being aged in ex-Bourbon casks and, post-blending, given time to marry in cask. There are some quite hefty pot-still aromas to start with: banana jam, coconut, maple syrup. This rich and creamy character carries on to the soft but chewy palate, which dries slightly into macadamia, dried apricot, tea-like tannins, honey, and stewing fruit.

We get lime zestiness with ginger beer, a decent, gentle fusion with the clementine juice, and a gentle, if slightly flabby, rum and cola. Coconut water edges ahead due to its deeper, more considered, quality; a great sundowner or after-dinner drink. The Old-Fashioned rounds everything out with oak, but has sufficient sweetness to balance. A rich drink with a pecan-like finish.

TASTING VALUES			
4	Coconut Water	3.5	Clementine Juice
3.5	Ginger Beer	3.5	Cola
4	Old–Fashioned		

THE RUMS: AGED ENGLISH-SPEAKING CARIBBEAN

ST NICHOLAS ABBEY
5-YEAR-OLD 40% ABV

St Nicholas Abbey, one of only three remaining Jacobean mansions in the western hemisphere, was built around 1650 by Lieutenant Colonel Benjamin Berringer. It was given to Sir John Gay Alleyne on October 19, 1746 as a wedding gift. The Cave family owned it from 1834 until 2006, when Larry and Anna Warren bought it and began restoring the property and – most importantly for us – starting rum production.

Early bottlings were distilled and blended by Richard Seale, but this rum is from the abbey's own cane (it's cane-syrup based) and distilled in its hybrid pot still. No sugar is added. A high-toned, aromatic lead-off moves into marzipan, peach, crisp oak, sweet cicely, green banana, and cream. With water, there's the warming aroma of ink. The palate is pure and clean, with a creamy flow that comes as much from the rum as the oak.

Bajan rums often hide their complexities under a seemingly understated surface, although they can be discovered with mixers – apart from cola in this case. Clementine juice adds length and class to things; ginger beer goes into fenugreek and spice, with a dry finish. Coconut water elevates the mix, showing a natural sweetness, perfume, and, again, dryness. The Old-Fashioned is light and elegant, reminding me of old rooms with daylight streaming through the windows.

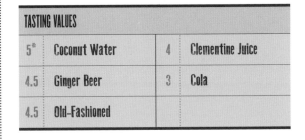

TASTING VALUES			
5*	Coconut Water	4	Clementine Juice
4.5	Ginger Beer	3	Cola
4.5	Old-Fashioned		

SMITH & CROSS 57% ABV

The cry went out from members of the new tiki generation: "We need the funk. Gotta have some funk!" They needed old-style Jamaican pot-still rums, but it was a style that had gone out of fashion. Enter Smith & Cross – or to be precise, enter Hampden Estate, which supplies the six-month-old, ester-rich, Wedderburn-style rum that gives pungency and weight, and the 18- to 36-month-old medium-bodied Plummer style that gives middle-range fruits.

The nose is a rich mix of mink oil and pineapple esters, then evolves into a mix of golden syrup and weirdly savoury decaying fruit, with jags of allspice and camphor. The palate is big, dry, and intense. The alcohol drives things forward, but the pot-still weight acts as a brake, allowing it to settle and coat the tongue, releasing a mix of roast chestnut and quince syrup tones.

The mixers flounder. Coconut water walks away in a huff; ginger beer makes things too fat and rich; cola gives a maraschino edge, but it's clearly just getting in the way. Clementine juice, though, is expansive, deep, and fascinating.

In the Old-Fashioned it remains true to itself, but stirred down there's more pineapple, banana, and a rich, soft, mellow mid-palate. It's a big-statement rum, but as George Clinton said, the funk is its own reward. Essential for punches and tiki drinks.

TASTING VALUES			
2	Coconut Water	4.5	Clementine Juice
3	Ginger Beer	3.5	Cola
4	Old-Fashioned		

ADMIRAL RODNEY
EXTRA OLD 40% ABV

Named after the British sailor who captured St Lucia from the French in 1726, this older-profile rum (the average is 10 years) is, perhaps surprisingly, a 100-per-cent column-still rum, but Saint Lucia Distillers has a trick up its sleeve. It uses its No 2 yeast, which is cultured from the yeasts on sugar cane and helps create a higher level of flavour compounds in the rum. The distillate is taken off at different points on the rectifying column, which further widens the flavour spectrum. The mature character is obvious on the start, with polish, rich wood tones, toasted nuts, then rhubarb jam and a background herbal element. The palate shows sweeter depths, with a more liberal (baked) fruitiness, then dried mango. It's complex, mellow, and layered with a firm tannic finish.

As this is a balanced, hefty rum with good solidity, should it be mixed? Yes: if you know where to go. Avoid cola unless you really like wood; ginger beer fails to excite, but coconut water instantly ups the distillery character, picking up the fresher elements and amplifying those top notes. A mass of tropical fruits greets you with the clementine juice: all energy and gentle sweetness. It's the Old-Fashioned, though, where this blossoms, as richer, earth tones expand alongside resin, cloves, and heavy florals. Layered, long, complex; highly recommended.

TASTING VALUES			
4.5	Coconut Water	4.5	Clementine Juice
3.5	Ginger Beer	2.5	Cola
5*	Old-Fashioned		

APPLETON ESTATE RARE BLEND 12-YEAR-OLD 43% ABV

The first record of rum being made at Appleton Estate is in 1749, although there is every chance that it was being distilled almost a century earlier. This makes it one of the great repositories of rum heritage, rich not only with the physical terroir of the surrounding area but the cultural terroir that comes with it. Rare Blend is where all of the estate's characters come together. It's a rum with depth and some pungency but all of the elements are controlled and balanced – impressive for a rum that is old by Caribbean standards and could easily be dipping into oakiness. The nose is funky, with rich, baked soft fruits, pot-still weight, and an added chocolate note. Water allows some kumquat to come out alongside mocha and crème brûlée. The palate is medium- to full-bodied with light grip and a ripe-fruit character with the muscularity of molasses, dried fruits, and bitter chocolate.

There aren't many rums that work across the board, but this does. Coconut water gets extra length, some toasty wood, and hillside funk; clementine juice picks up the earth tones as the molasses comes through. Ginger beer has weight and an extra dimension and a balanced finish, while cola deepens the molasses on the mid-palate. In Old-Fashioned world there's an upping of oak, and the release of elegant mature notes: leather and overripe fruit. A drink not to be taken slowly.

TASTING VALUES			
4.5	Coconut Water	5	Clementine Juice
4.5	Ginger Beer	4.5	Cola
5	Old-Fashioned		

CARONI 15-YEAR-OLD, VELIER
52% ABV

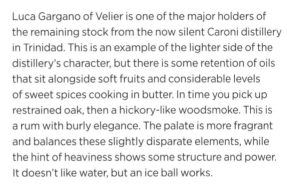

Luca Gargano of Velier is one of the major holders of the remaining stock from the now silent *Caroni* distillery in Trinidad. This is an example of the lighter side of the distillery's character, but there is some retention of oils that sit alongside soft fruits and considerable levels of sweet spices cooking in butter. In time you pick up restrained oak, then a hickory-like woodsmoke. This is a rum with burly elegance. The palate is more fragrant and balances these slightly disparate elements, while the hint of heaviness shows some structure and power. It doesn't like water, but an ice ball works.

Mixing is like observing a speed-dating event, with Caroni as the poor schmuck in the corner no one seems to want to go out with. There's light oak with coconut water, but it's not an alliance. Cola promises a lot – there are blackcurrants galore on the nose, but no harmony on the palate. Ginger beer is minty but lacks length. The last person to visit Caroni's table is clementine, and here things finally work: almond, juicy fruits, pineapple chunks... it's just lovely. (You see? There's always someone out there for you.)

As an Old-Fashioned, the rum becomes unusually shy to start with, then releases a burst of varnish, so leave it on its own – or with its darling clementine.

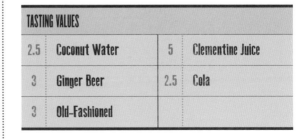

TASTING VALUES			
2.5	Coconut Water	5	Clementine Juice
3	Ginger Beer	2.5	Cola
3	Old-Fashioned		

CARONI 1999 (BOTTLED 2015), RUM NATION 58% ABV

Welcome to the Trini style of funk. This is very Caroni in its nose of furniture polish and freshly dubbined boots, before a note of hospital corridors takes you back toward Islay. On top of these weird oils and phenols are raisin cookies, campfires, and treacle scones. A weighty glass of rum, it then becomes smokier with water. This is one for confirmed Caroni-lovers, though maybe not for beginners.

The question is: will any mixer work here? The simple answer is no. Coconut water brings out a nose that's not unlike a petrol station forecourt at 3am; clementine juice hides the rum; while the ginger beer is akin to a sweaty saddle – though some people do like that aroma. Cola is the best of the bunch, even if it does take you on a high-speed trip on the back of an old Harley-Davidson into a wet jungle.

Craft an Old-Fashioned from it and you will increase the oils but also produce a surprising hint of strawberry alongside antiseptic cream. You're back in the hospital (maybe you crashed the bike?). Next time, just have it on its own.

TASTING VALUES			
2	Coconut Water	2	Clementine Juice
2	Ginger Beer	2.5	Cola
2.5	Old-Fashioned		

DOORLY'S 12-YEAR-OLD

40% ABV

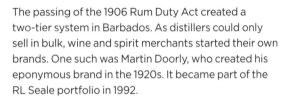

The passing of the 1906 Rum Duty Act created a two-tier system in Barbados. As distillers could only sell in bulk, wine and spirit merchants started their own brands. One such was Martin Doorly, who created his eponymous brand in the 1920s. It became part of the RL Seale portfolio in 1992.

This is a good example of Richard Seale's thinking in terms of wood. A blend of pot and column rums, 90 per cent is aged for 12 years in ex-Bourbon casks, while the remaining percentage is matured for the same time in Madeira casks. The Madeira has given this a very slightly red tint and you immediately pick up sweet dried fruit, nutty oxidized notes, marmalade, vanilla, and molasses. In time, classic mature notes of cedar and leather from the pot still emerge. The palate is balanced, dense, and soft, gaining focus in the middle, where deep, raisined fruits lurk. It's rich rather than sweet; robust, not heavy.

There's an element of inconsistency when it comes to mixing. Ginger beer adds nothing. Clementine juice, though pulling out green mango and some dryness on this finish, is so-so; but coconut water helps to amplify the elegance of the rum. Cola shifts into rum-and-raisin and is pretty satisfying. That said, old rums need Old-Fashioned treatment. Here, the dried fruit is the balancing agent and there are new notes of cherry pie emerging in its supple, rippling, length. Highly recommended.

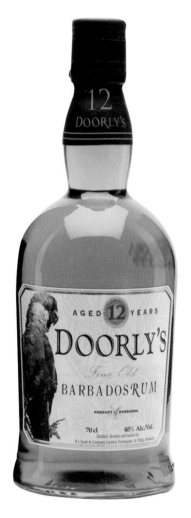

TASTING VALUES			
4	Coconut Water	3	Clementine Juice
2.5	Ginger Beer	4	Cola
5*	Old-Fashioned		

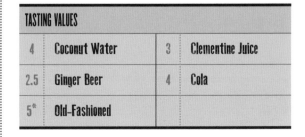

EL DORADO 12-YEAR-OLD
40% ABV

The astonishing collection of stills at the Diamond Distillery is testament to Guyanese distillers' understanding that the nature of the variety of marks the country produced was down to the stills. Today, the distillery has 10 stills: a double pot (Port Mourant), single wooden pot (Versailles), a wooden Coffey (Enmore), two Savalle columns (Uitvlugt), three metal Coffey stills, one double column, and a five-column set-up.

Each of its rums contains a different blend of the marks made on each still. This 12-year-old is a blend of marks from the "modern" Diamond Coffey stills and heavy rum from the Port Mourant wooden pot still. There are scented, violet-like elements here that move into briar fruits. The impression is of a rich, mellow, massing of fruits before some coconut oil and treacle emerge. The palate is big, sweet, and layered, with tones of mulberry jam, liquorice, black cardamom, pepper, soft fruits, and currant. It then develops a slightly bitter edge. It's old-style, sweet Demerara rum.

Cola makes for a flabby combination. Clementine juice, though, adds a pleasing bite and lift, while with ginger beer the pot-still element comes forward, adding richness to the mid-palate and cleaning up the finish. Coconut water builds in complexity and brings a drier balance. The Old-Fashioned has huge impact, but fails to showcase the complexity that lurks in the rum.

TASTING VALUES				
5	Coconut Water		4	Clementine Juice
4	Ginger Beer		2	Cola
3	Old-Fashioned			

EL DORADO 15-YEAR-OLD
43% ABV

It's fair to say this was the brand that kick-started the new wave of premium rums. Demerara Distillers' visionary chairman, Yesu Persaud, felt that there was an alternative to being solely reliant on high-volume, low-profit bulk exports. The fact that he decided to back this belief with a 15-year-old rum – an age statement unheard of in the 1990s – made this a bold statement of intent.

This is a blend of marks from the Enmore wooden Coffey still, the metal Coffey stills, Port Mourant double wooden pot still, and the Versailles single pot still. It's the pots that dominate the nose, with coffee notes and a mix of black banana, PX Sherry, and a tiny whiff of magic marker, before molasses mixed with damp earth after a tropical storm emerges, with a smoky edge. On the palate there's a shift to liquorice halfway through, along with subtle wood. It's distillate rather than oak-driven and finishes (very) sweetly.

There's a medicinal element to the coconut water and the feeling that rum is only tolerating the mixer out of good manners. Clementine juice, amazingly, loses the rum in its depths, while ginger beer is slightly medicinal. Though cola shows the black-banana element, there's a loss of balance in the middle as things get sweet. Rum-soaked timbers come through in the Old-Fashioned, which is all plums and blackcurrant. It has weight and depth, but the sweetness is too much. Have it with ice.

TASTING VALUES			
3	Coconut Water	3	Clementine Juice
3	Ginger Beer	2	Cola
3.5	Old-Fashioned		

THE RUMS: AGED ENGLISH-SPEAKING CARIBBEAN

MEZAN XO JAMAICA 40% ABV

The concept behind the Mezan range is to show "untouched" rums: no caramel tinting, flavourings, sugar, or chill-filtering. They are mostly single-estate rums, this being the exception, which is a blend of Worthy Park and Monymusk.

Very pale in colour and showing the use of refill casks, there's full-on Jamaican drive from the off here, with smoky phenols, pineapple, light oils, and lots of lemon (in a similar vein to some Long Pond distillates). The pot-still power then begins to build, adding energy and pungency that lifts the aroma toward acetone, but stops just before: running past a nail bar rather than sitting in one. The palate is clean after a light start, with kiwi fruit, molasses, and good acidity. The pineapple (now roasted) re-emerges, and with water, some creamy oak.

The combination of low oak and punchy Jamaican pot still means the rum is in charge with coconut water; any attempt to persuade things to harmonize is swatted aside. Cola is a waste of effort, but ginger beer works well because there's no oak in the way, which allows it a clear run along the palate, adding zest and linking with the lemon notes. Not for the first time, clementine juice is the most successful mixer, with its fruit flowing into the pineapple and really emphasizing the banana tones, making this an impressive (and presumably potassium-rich) energy drink. The Old-Fashioned has weight, but the rum lacks the extra dimension given by more active casks. If in doubt, sip it on its own.

TASTING VALUES			
3	Coconut Water	4	Clementine Juice
4	Ginger Beer	3	Cola
3	Old-Fashioned		

MOUNT GAY XO 43% ABV

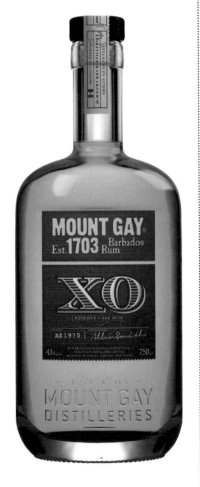

An older age profile, this is Mount Gay in contemplative mood; the rum's components are between eight and 15 years old, with significantly more pot-still input. The nose is oak-driven, firm, with a Mozart-like (furfural) note that drifts into cigar humidor, dried fruits, light raisin, and an old leather armchair. All the way, though, there's this molasses undertow. The palate is more lifted, with Mount Gay's note of honeyed citrus before maraschino and roasted coconut break through. Water backs up this feeling of structure. It's a good after-dinner rum that begs for a Romeo y Julieta cigar in the hand.

As you move up in terms of age and structure, so the interaction between rum and mixer changes. With light, fresh rums the mixer helps to reveal complexities. Once the rum has settled into its mature period, oak is more prominent, then more barriers are erected. Here, with the clementine juice, there's depth and a slightly drying offset. Ginger beer is not exactly subtle, with all the elements turned up rather than harmonizing; cola pulls out that leathery note, and coconut water brings an almost sappy element.

The Old-Fashioned is, on balance, the best route to take. Here is a herbal, spiced element, rose petal, citrus, and a pleasant bittersweet note that plays off the central sweetness. Rather lovely, in fact.

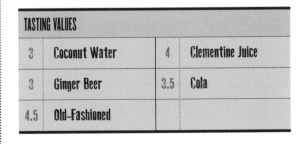

TASTING VALUES			
3	Coconut Water	4	Clementine Juice
3	Ginger Beer	3.5	Cola
4.5	Old-Fashioned		

THE RUMS: AGED ENGLISH-SPEAKING CARIBBEAN

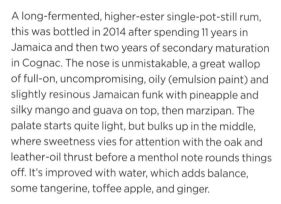

PLANTATION RUM
JAMAICA 2001 42% ABV

A long-fermented, higher-ester single-pot-still rum, this was bottled in 2014 after spending 11 years in Jamaica and then two years of secondary maturation in Cognac. The nose is unmistakable, a great wallop of full-on, uncompromising, oily (emulsion paint) and slightly resinous Jamaican funk with pineapple and silky mango and guava on top, then marzipan. The palate starts quite light, but bulks up in the middle, where sweetness vies for attention with the oak and leather-oil thrust before a menthol note rounds things off. It's improved with water, which adds balance, some tangerine, toffee apple, and ginger.

Cola has a slightly off-putting oiliness, ginger beer gives a spicier, livelier take on funk, while clementine juice helps to emphasize the fruit, and while it's short, it's rather fine. Coconut water moves things back to Jamaica with masses of bananas and high molasses.

It's hard for anything to hide when made into an Old-Fashioned; it reveals flaws, but also pulls out new flavours, such as the smoky elements that emerge here. The base sweetness of the rum is also enhanced. On balance, however, I think that the best way to enjoy this is to pour a glass, add a cube of ice – and light a cigar, if you fancy it.

TASTING VALUES			
3	Coconut Water	4	Clementine Juice
3.5	Ginger Beer	2.5	Cola
3	Old-Fashioned		

THE RUMS: AGED ENGLISH-SPEAKING CARIBBEAN

PLANTATION XO
20TH ANNIVERSARY 40% ABV

Plantation Rum is the brainchild of the enterprising Alexandre Gabriel, who not only heads Pierre Ferrand Cognac, but makes Citadelle Gin, triple sec, and a host of other spirits. He takes a Cognaçais approach to maturation by bringing American oak-matured stock from the Caribbean to Cognac, where it has a period of secondary maturation in small casks. The XO was created to celebrate Alexandre's 20th year in charge of Ferrand. It's a blend of Barbadian pot- and column-still rums and spends 18 months in Cognac casks. The nose is thick, with lightly smoky elements and a resinous note with a hint of blue cheese – rum *rancio*? This rich quality continues with notes of five spice, suntan lotion, then galangal and rose. The palate is huge and quite sweet, with coconut, apricot, and mango notes, a hint of Falernum, then blackcurrants. As it opens it gets richer and darker.

There's a heavy sweetness unusual for Bajan rums, which dents the cola mix and dominates the ginger beer. The clementine juice induces a woody toastiness that disturbs the flow. Coconut water is a natural ally, though the mix does become very Bounty bar. You're on safer ground with the Old-Fashioned, where citrus adds top notes. Everything then becomes better balanced, with sweet, fruity, bitter, and spicy elements working together in a dazzling fashion. Alternatively, have it with a cigar.

TASTING VALUES			
4	Coconut Water	3.5	Clementine Juice
3	Ginger Beer	3	Cola
5	Old-Fashioned		

RL SEALE'S 10-YEAR-OLD
43% ABV

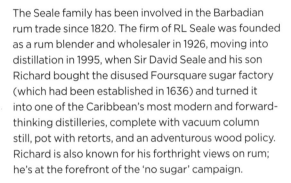

The Seale family has been involved in the Barbadian rum trade since 1820. The firm of RL Seale was founded as a rum blender and wholesaler in 1926, moving into distillation in 1995, when Sir David Seale and his son Richard bought the disused Foursquare sugar factory (which had been established in 1636) and turned it into one of the Caribbean's most modern and forward-thinking distilleries, complete with vacuum column still, pot with retorts, and an adventurous wood policy. Richard is also known for his forthright views on rum; he's at the forefront of the 'no sugar' campaign.

Here, the nose has Seville orange and juicy tropical fruits to the fore, and while on the lighter side, it has a distinctly firm frame (sugar is not used to disguise grip). With water, there's some toasted hazelnut. The palate is clean and concentrated on the start, with plenty of sweet spice and apple. Water helps it move into heavier, honeyed depths, with the citrus peels adding a frisson. This is a rum suited to single-malt drinkers.

Mixing here is little more than an academic exercise. Ginger beer seems light, cola clashes, and clementine juice is diminished. Even coconut water makes something that's little more than an easy-going drink. Head into Old-Fashioned territory, however, and a slow and steady journey of exotic flavours and textures is laid out in front of you, from sweet to rich, spiced to exotic.

TASTING VALUES			
3	Coconut Water	3	Clementine Juice
2.5	Ginger Beer	2	Cola
5	Old-Fashioned		

RHUM AGRICOLE, FRENCH DÉPARTEMENTS & HAITI

People raised on molasses-based rums find it hard to comprehend cane-juice rums. The aromas are different, the structure is finer, the rums (or *rhums*) are dry. Equally, it's hard for those used to cane-juice rums to find their way into molasses-based spirits. Where is the green quality? Why is there no mineral/marine note? How come the flowers are not so apparent? And what's all this sweetness?

Agricole isn't going to change, so those on the molasses side have to train their palates to think of it differently: a new world of flavour. Embrace the vegetal notes, the heightened spice, the red fruits, the saline quality; revel in the fact that the aged examples have fine tannins and clearly defined structure. These are world-class spirits.

Ironically, these are *rhums* that are most likely to appeal to drinkers coming to the category from single-malt Scotch; they have grassiness, spice, and balanced (but apparent) oak. They also speak of their terroir: the vintage conditions, the location of the cane fields, the type of cane, the subtle differences in still design.

There are outliers here – two molasses-based French *rhums*, and three from Haiti, two of which are aged and aligned to *agricole*. The other is a *clairin*, which takes rum/*rhum* into a new world and nods toward mezcal. Who said rum was easy?

Ti Punch is the natural cocktail of choice for *rhum agricole*, and is what I chose to try here.

J BALLY RHUM BLANC 50% ABV

Although the Bally Distillery closed in 1989, today the brand is distilled at Martinique's St James distillery in a Creole column still. It is slightly fatter than many of the island's other blancs, with more roundness on the nose and a warm, earthy quality with some firmness behind. It needs some time in the glass to begin to open into Williams pear and a little apricot skin.

The palate has a slightly marine quality to it, which (weirdly) brings to mind freshly shelled prawns. Things begin to become more land-based by the mid-palate, which suggests soft and quite yielding melon fruits, then that humid-earth note seen on the nose. When water is added, things become more straightforward. Ultimately it remains big, but firm and dry and slightly tense at the end – indicative of a *rhum* that is more relaxed in an aged setting.

As a Ti Punch, however, it becomes considerably more expressive – almost delicate, in fact, as the citrus adds some lifted top notes and more of a bracing cut-through of acidity. Now, some floral top notes are released over this mineral core.

TASTING VALUES		
3	Ti Punch	

THE RUMS: *RHUM AGRICOLE,* FRENCH *DÉPARTEMENTS &* HAITI

CLAIRIN SAJOUS 51% ABV

It was long believed that Haiti only had one *rhum* distillery. In reality, it has around 500, virtually all of which make a style called *clairin*, which is to rum/*rhum* what mezcal is to tequila. You also need to approach it in the same way: on its own terms rather than in reference to other cane-based spirits. The cane used is mostly organic, hand cut, wild yeast-fermented, and distilled in a variety of types of stills, mostly pots some with small rectifying columns. *Clairin* has retained its links to land and to society. It's a medicine, a ritual spirit, as well as a drink. Its "discovery" sadly typifies the West's attitude to Haiti: misunderstood, ignored, and deliberately impoverished.

The difference between all the examples lies in the cane used, the nature of the wild yeasts, and the skill of the distiller. Michel Sajous' distillery is in St-Michel-de-l'Attalaye, in the middle of a 30-hectare plantation of different varieties of cane, prominent among which is the *cristalle*, the last *canne à bouche* allowed in the AOC Martinique Rhum Agricole. It has been double distilled in pots to 53.5% ABV.

The nose is pungent, bone-dry, and intense with a herbal quality and a little touch of putty. The palate takes you back to an earthy element, with pure, sweet fruits, rotting flowers, grass, angelica, tarragon, and banana. Water brings out the fruits. It's simultaneously old and new. Drink it straight.

TASTING VALUES		
N/A	Ti Punch	

CLÉMENT CANNE BLEUE 2013 50% ABV

This single-varietal *rhum blanc* – the first to be designated as such – was launched by Clément in 2001. Like all *rhum agricole blanc* it is rested in stainless steel for a period to allow the more volatile elements to be vented off.

The nose is heady, dry, and quite intense, with light vegetal notes and a slight dusty/biscuity tone in the background. Ultimately, though, it is delicate and precise – particularly so when you add some water, when it becomes more floral and herbaceous, with good complexity and touches of violet.

The palate is clean, with crisp pear, green bean (tank-fermented Sauvignon Blanc), yellow fruits, and apple core flavours. Dilution brings back the off-dry quality and some fresh spices.

Mixed as a Ti Punch, there is a great floral lift-off, with the lime oils cutting through the vegetal aromas. The full character has been retained, but the touch of sugar in the drink helps to add a little weight and some texture to the mid-palate and brings out a little hint of fennel seed at the end.

TASTING VALUES		
4.5	Ti Punch	

CLÉMENT PREMIÈRE CANNE
40% ABV

A mix of different varieties of cane are used on this *rhum blanc*, which is aimed primarily at the bar trade. It has been rested for nine months in tank being slowly reduced in strength to 40% ABV – lower than most *rhum agricole blanc*.

The nose is filled with the anise edge of fresh basil and a little touch of white chocolate followed by the mineral/marine touch that appears in some white *rhums*. It is less overtly vegetal than the Canne Bleue (*see opposite*), being sweeter and fatter, with touches of lemon sherbet when water is add.

The palate is medium-sweet and clean, with some chalkiness and light-green fruits. Now more of the grassy, cane-leaf notes begin to emerge, and though the impact is a little light for those used to the full-on, uncompromising hit of traditional examples, this lower alcohol will make things more approachable for anyone new to the *agricole* game, but without ever losing character. A pure, distillate-driven *rhum*.

That mineral quality comes through more strongly as a Ti Punch; this is cool and clean, with the lime cutting through, making everything redolent of apricot, tomato leaf, and strawberry. The amount of sugar that you add could be an issue here because it is very easy-drinking – so be careful.

TASTING VALUES		
4	Ti Punch	

THE RUMS: *RHUM AGRICOLE, FRENCH DÉPARTEMENTS & HAITI*

KARUKERA RHUM BLANC
50% ABV

The oldest distillery on Guadeloupe, located in the heart of the Domaine du Marquisat de Sainte-Marie, the Karukera *rhums* are made from cane grown on the 28 hectares that surround the company. For this white, *canne bleue* ("blue cane") is used.

It has a fresh cane nose with a firm backbone, some tropical notes, soft citrus, and some sweetness, giving a gentle, but lightly complex start. Its slightly broader qualities begin to develop slowly and things start to become more pungent before dipping into overripe fruitiness. With water there is more of a powdery quality – hot, dusty tracks on a hot day.

The palate is quite different: medium-dry with a slightly earthier base note, touches of wild tarragon, and fennel. Water helps to show that this is a really solid, well-made *agricole* with some finesse.

As a Ti Punch it shows vegetal notes that hint at acetone-like levels of intensity backed with vine flower and pea shoots. It manages to mix the two side of *agricole*: the fresh, green cane and the fleshy fruits. Well balanced too.

TASTING VALUES		
4	Ti Punch	

NEISSON RHUM BLANC
52.5% ABV

Family-run Neisson is the smallest producer in Martinique and uses cane grown on 34 hectares between Le Carvet and Saint-Pierre on the drier and hotter Caribbean coast in the northwest of the island. Founded in 1931, it is now run by Gregory Vernant-Neisson. The firm distils in a copper Savalle still and ferments for longer than most estates (between 72 and 96 hours), upping ester levels. This rum is made from *canne bleue* ("blue cane"), all of which is grown in Thieubert beside the distillery and close to the sea.

The nose is huge: stronger than most, and intense, with a rich, almost oily, fleshy weight that takes you into super-ripe fruits and botanical garden hothouses. This is backed with pomegranate, some raspberry (leaf and fruit), and then tinned peaches.

This perfumed quality continues on the palate with a little jag of citrus before moving into green, almost mossy notes. It's quite explosive, with a dusty element and a marine note, complex and rich.

As a Ti Punch you get a big, pungent hit, then persimmon, and tomato-leaf funkiness. It needs a touch more sugar to balance the mix of vegetal tones, huge floral hints, liquorice, and a perfumed, oily quality.

TASTING VALUES		
4.5	Ti Punch	

THE RUMS: *RHUM AGRICOLE, FRENCH DÉPARTEMENTS & HAITI*

RHUM JM BLANC 50% ABV

The JM distillery is located in the far northwest of Martinique, on the slopes of Mount Pelée, surrounded by cane fields and pineapple and banana plantations. This *rhum* is distilled in Creole column to a strength of 72% ABV and then rested for four months in stainless-steel tanks.

The nose is spicy and layered, with fruits (ripe and green) rather than vegetal notes. In time, there is a hint of papaya and whiffs of banana from the nearby groves before green apple makes an appearance, alongside fresh cane, then the distillery's signature note of white pepper.

The palate again shows this light mineral quality alongside some plump ripeness and, with water, more textural oiliness. The palate is dry and full, coating the mouth, with those peppercorns adding a catch toward the end alongside fresh red berries. A powerful *blanc*, with the most obvious structure.

Things start in a slightly chalky fashion when made into a Ti Punch, but there's a supple quality to this drink, which delivers surprising delicacy in the middle. The lime is the key here, stirring those red fruits, citrus, and pepper notes into a complex and elegant finish.

TASTING VALUES		
5*	Ti Punch	

RHUM RHUM PMG 56% ABV

The brainchild of Luca Gargano of Velier and the presiding genius of fruit-spirit distillation Gianni Capovilla, this *blanc* is distilled at the Bielle Distillery on Marie-Galante, but in a very different manner to all the other local *rhums*. A selection of local canes is used, the most significant being the romantically named B.47.258 – aka canne rouge.

Fresh cane juice is fermented at 20–22°C (68–72°F) in temperature-controlled vats for seven to nine days. It is then double-distilled in pots fitted with bain-maries rather than a single column. Finally, it is rested for a year in steel before bottling.

The nose is perfumed, with pineapple elements, some rich oils, dried pears, and raspberry. Rather than being vegetal, it has this exotic fruit note that only hints at cut cane, green grass, and a zesty, grapefruit-like citric element. In time, the mineral/seashore aromas begin to come across – almost like green olives in brine.

The palate is a paradoxical mix of concentrated fruit with a bone-dry spine. Water is needed to fill it out. There is always light, edgy alcohol.

Ti Punch both expands in terms of complexity and knocks back some of the more outré aromas. It becomes heavier, more complex, with a little dried rose petal. It must be served ice-cold.

TASTING VALUES		
5	Ti Punch	

SAVANNA LONTAN GRAND ARÔME 40% ABV

The Savanna Distillery on the island of La Réunion was founded in 1948, and has been distilling at its Bois-Rouge site since 1992. The distillery uses cane juice, and molasses distilled in a selection of column stills.

This white is made from molasses, which are given a very long ferment to make a style known as *grand arôme*. This is often translated as "high ester", but the character here is less of the pungent pineapple/glue notes of Jamaica's more extreme examples. Rather, the nose has a powerful and deep aromatic element. The fruits are to the fore: rich, ripe papaya and canteloupe, then banana, redcurrant, and mirabelle. When it opens, there's a background note of molasses.

The palate starts quite fragrant and with surprising heat for the strength. This has real verve and character.

It might seem to be a less-than-orthodox *rhum* to use for a Ti Punch, but I went with it for consistency's sake – and out of sheer curiosity. Here, the fruits really come through, but it's the way the lime oils combine that settle this down into a quite a rich, late-afternoon drink.

TASTING VALUES		
4.5	Ti Punch	

J BALLY RHUM AMBRÉ 45% ABV

The Bally estate (originally called Lajus) started in 1690, but was one of the many plantations devastated by the eruption of Mont Pelée in 1902. It was bought by Jacques Bally soon afterward, and switched entirely to *rhum* production. Today, all *rhums* are made at the St James Distillery.

A fresh and relatively young *rhum* – it has been in cask for two years – the nose has the Bally weight to it: fatter and slightly more humid in style, but now with a drier element, suggestive of French oak being used: that hint of cardamom and dry-roasted spices. It starts with lightly oxidized almost cheesy notes, dry oak, some nuts, and still the remnants of the vegetal aspects of the straight distillate reminiscent of flower stems in water, wet plaster, then green peach.

The plate is leaner, direct, and quite hot, with a firm structure and more of an earthy, herbal quality than the *blanc* (*see* p.140), though the grassy cane element can be seen. Water, however, allows some more caramelized elements to come through.

As a Ti Punch it retains the purity of character: clean, with decent depth, deepening into sugar before the dusty spices take over on the finish. Young and vibrant, but good.

TASTING VALUES		
3.5	Ti Punch	

BIELLE RHUM VIEUX (HORS-D'AGE) 42% ABV

Based on the island of Marie-Galante, what is now the Bielle plantation started growing coffee in 1769, with sugar production commencing only in 1826. The distillery has been in production since the latter part of the nineteenth century, becoming specialized as the small *sucrottes* closed down with the advent of beet-sugar production in Europe.

This *vieux* is aged for four years in a mix of ex-Bourbon and then ex-Cognac casks. Light amber in colour, the aroma is quite delicate and refined, which is very different and more fine-boned than Martinique. It is more floral and, dare I say it, more Cognac-like, with sweet spices and a whiff of new suede shoes, then lemon balm and lemon meringue pie.

The palate is sweet and fresh, with some clean floral notes mixed with vanilla and light tobacco from the oak; energetic, but with fine-grained tannin. A charming and pretty *rhum*.

When mixed as a Ti Punch, the cane juice really begins to come through. The aroma is like walking into the distillery as the cane is being crushed, giving off this sweet, heady (but not vegetal) aroma. I went light on the sugar as the *rhum* has sufficient inbuilt sweetness to carry. It is equally good on its own.

TASTING VALUES		
3.5	Ti Punch	

KARUKERA RHUM VIEUX RÉSERVE SPÉCIALE 42% ABV

The Guadeloupe-based Karukera (the original name for the island) distillery is following a growing trend within the *agricole* world of using ex-Bourbon casks for maturation rather than the more traditional ex-Cognac. This not only gives the *rhum* more of a sweet, vanilla/coconut element, but also imparts lower tannins and brings the whole package more in line with a more familiar aromatic profile.

This example has spent four years in ex-Bourbon and shows a big, plump nose with fresh plums, liquorice, and the start of a leathery, rummy *rancio* element, which is surprising at this age and might be indicative of older stock being blended in, or hotter warehousing conditions. Whisky- and Cognac-lovers apply here. There are some nutmeg and rooty notes, then stewing pear and a waxy element. The palate has real attack, with touches of long pepper, cardamom-roasted hazelnuts, and sweet fruits. Then it lengthens into aromas akin to a second-hand bookshop.

As a Ti Punch this is a little indistinct, with a light caramel note and becomes a little too limey to be perfectly balanced. Best with ice.

TASTING VALUES		
3.5	Ti Punch	

RHUM JM XO (BOTTLED 2014) 43% ABV

At approximately six years of age, this is mid-range in terms of maturity for the estate. At first if taken blind you think "Calvados?" because there is this very pure, clean apple and pear note that comes through, to which is then added spice, canvas, and anise. The fruits become longer and riper, with a pleasant laurel/geranium character. With water it remains subtle, reminiscent of a classic gentleman's cologne, and slightly powdery.

On the palate, the pepper sits alongside more intense spices and a little ginger. Although it seems to ripen in the middle, there is always a prickly, peppery heat and a certain element of tension, as if all the aromatics were not quite fully worked out.

As a Ti Punch there is more banana, but a dryness still dominates. This is structured and quite elegant, with some sweet vanilla and gentle oils as the fruits fold themselves into the mix and everything, finally, coheres into something greater than the sum of its parts.

TASTING VALUES		
5*	Ti Punch	

SAVANNA CUVÉE SPÉCIALE 5-YEAR-OLD 43% ABV

Made on La Réunion, this molasses-based *rhum* is a blend of Savanna's *traditionelle* and *grand arôme* styles – the latter the product of a two-week fermentation – aged in ex-Cognac barriques. They are older barriques, according to the nose, as the ingress of the oak is very gentle and well balanced. The aroma shows top notes of frangipani followed by guava and passion-fruit accents before the fresh date notes suggestive of age come through, along with a little touch of vanilla, white chocolate, and some oils.

The palate is considerably spicier, with some cumin and green cardamom, then fenugreek seed, which moves toward molasses – itself held in check by some honey. The finish shows spiciness, which drifts toward the mineral side.

Balanced and complex, this is a graceful molasses *rhum* that deserves to be better known.

A versatile spirit for all mixed drinks, this works equally well as a refined Old-Fashioned or as a slightly more spicy and long Ti Punch.

TASTING VALUES		
5	Ti Punch	

THE RUMS: *RHUM AGRICOLE,* FRENCH *DÉPARTEMENTS* & HAITI

TROIS RIVIÈRES VSOP RÉSERVE SPÉCIALE 40% ABV

The Trois Rivières estate is based in Ste-Luce in the southwest coast of Martinique. Originally a large sugar plantation with distillery attached, it has specialized in the production of *rhum* since 1905, starting with equipment from the Dizac Distillery in Le Diamant. A second column was installed in 1980. In 2004, the production facility moved to Rivière-Pilote but with the original column still intact.

This VSOP has been aged for an average of five years in French oak casks. The nose is typical of how the mature nose of *agricole* moves from the vegetal side to one where fruits and spices have more of a say, but without losing the fine, dry, element seen in *blanc*. There are touches of coconut and vanilla here that drift toward nutmeg and an almost chocolate, rye-like intensity, before water brings out fresh apricot.

The palate is relatively discreet, with touches of syrup and oak-driven flavours: vanilla, toast, a hint of smoke, cinnamon, and more of the menthol pepperiness. If you like Four Roses Bourbon, you'll love this.

As a Ti Punch it has a lemon-peel zing, which softens into quite a luscious palate, with the wood well harmonized within the whole.

TASTING VALUES		
4	Ti Punch	

THE RUMS: *RHUM AGRICOLE, FRENCH DÉPARTEMENTS & HAITI*

J BALLY 2000 43% ABV

Bally is another slow-maturing house, and its vintage releases are where – for me at least – it performs the best. This is a classic mix of the aromas of dusty roads hung with the oppressive heat of tropical blossoms after a rainstorm.

The house style is very honest in its direct, classical, approach to things. There's oak, there's roast chestnut, there's a sense of concentration and age. The palate is filled with orange to start with, then comes some grip and a sensation of flowers lying in the dust. It offers a lovely, slightly old-fashioned palate with low tannins that becomes smooth and sweeter than the nose suggests, with the sense of fresh carpentry that moves then into pine sap.

Bally's dryness then helps as a Ti Punch, as its firmer elements act as a counterpoint to the sugar and is the serve I think works best. It's very *agricole* – and that's what Bally does best.

TASTING VALUES		
4.5	Ti Punch	

BARBANCOURT RÉSERVE SPÉCIALE ★★★★★ 8-YEAR-OLD 43% ABV

The legendary Haitian distillery of Barbancourt was founded in 1862, in Le Chemin des Dalles in Port au Prince, by Dupré Barbancourt. Originally from the Charente region of France, he adapted Cognac methods to produce his rum. Ownership passed through his wife to his nephew Paul Gardère, and in time to his son Jean, who moved the distillery to the family's domaine and expanded production. It is now run by Thierry Gardère. Barbancourt employs 250 people and is actively involved in a large number of nonprofit initiatives, including education, health, sports, and the arts.

Production is primarily from cane juice, though at the end of crop this can occasionally be supplemented by some cane syrup. Barbancourt speaks of using the *Charentaise* method for distilling, but the first distillation is in a copper column, giving an initial distillate (*clairin*) of 70% ABV, the second in a pot still, giving a final distillate strength of 90% ABV. Maturation is in French oak.

The elegant nose shows spiced tones and some oxidized elements, fresh and baked fleshy yellow fruits, and artichoke. With water there's a hint of herbaceousness. The palate is medium-bodied, with tangerine, fig, cracked pepper, and understated wood. This has real elegance. It is at its best as a sipper, but makes a gentle Ti Punch, where the softness and spice combine well.

TASTING VALUES		
4	Ti Punch	

THE RUMS: *RHUM AGRICOLE, FRENCH DÉPARTEMENTS & HAITI*

CUVÉE HOMÈRE CLÉMENT, HORS D'AGE 44% ABV

Homère Clément, the pioneer of *rhum agricole*, bought Domaine de l'Acajou in the southeastern part of Martinique, in 1887. Now renamed Habitation Clément, it remains a tourist attraction, with gardens, gallery, and plantation house. Distillation now takes place at the nearby Simon plant.

This premium bottling has been aged in a mix of ex-Limousin barriques and recharred ex-Bourbon barrels and is a blend of *rhums* from different years, the oldest being 15 years old – a fair age in this tropical climate.

There is a huge amount of orange peel and clove to start with, which is strangely reminiscent of a tropical Glenmorangie. It is scented, supple, and soft, with mixed peels, passion-fruit, mango, and gentle tannins. The citrus recedes with water, seemingly now absorbed. The palate shows it to be refined and not as perfumed as the nose suggests, while with water there is anise and that citrus note now: more pink grapefruit and lemon switched to the back. With water it is somehow deadly serious but frivolous at the same time.

As expected, the citrus is amplified when made into a Ti Punch, with more kumquat mixing with a previously obscured green herbal element, lime, and dusty spices. Subtly complex, gentle, and ultimately a highly rewarding drink – and a signature serve.

TASTING VALUES		
5	Ti Punch	

NEISSON 2004 SINGLE CASK (BOTTLED 2015) 45.4% ABV

For this single-cask bottling, canes from Martinique's Godinot plateau were used exclusively. Ageing took place in ex-Bourbon casks (a shift from Neisson's more common 30 per cent new French oak). The nose is big and mature: evidence of what is prolonged maturation in this climate, even with regular consolidation of the casks. The terroir, of course, has an impact. The northwest coast is dry and very sunny, which does help drive a more rapid maturation cycle.

It could almost be an Amontillado Sherry in its touches of oxidized nuttiness, but things are brushed with some cherry pipe tobacco, cedar, and lilies, as well as the red fruits seen in the *blanc* still in evidence – albeit now cooked. A light mineral, marine element is followed by powdered rose petal hints.

The palate is equally aromatic and dense but manages to have lift, balance, and complexity. Spices kick off, adding a crisp, sweet, roasted element without ever overtaking the unctuous depths. It's *agricole* on the dark side.

Ti Punch-wise, things become more oak-driven and accordingly spicier, as well with lingering complexity. This is a big, contemplative, and excellent drink that works equally well as an Old-Fashioned (as do many Bourbon-matured *agricoles*).

TASTING VALUES		
5	Ti Punch	

RESERVE RUM OF HAITI, DISTILLED 2004, BRISTOL CLASSIC RUM 43% ABV

Distilled at Barbancourt, this is an example of an "Early-Landed" style most commonly seen in Cognac. The term refers to the fact that the distillate is taken from its place of origin and matured in Britain – meaning it arrives earlier than is normal.

Bristol Spirits' (very) cool, damp warehouse at Wickwar, Gloucestershire, provides a completely different set of conditions, slowing the maturation down, and producing a rum that is more delicate and fine, with less oak impact.

The colour is, as you'd expect, pale for its age, and the nose has touches of lint, rose, and lime pickle, and remains distillate rather than oak-driven. It offers a hint of banana chew, more green cane/olive than is standard with Barbancourt, and a mildly phenolic Haitian funk element that builds on the palate.

This element is mixed there with fruit syrups, more banana, and pepper-coated cashews. A relaxed and fascinating *rhum* and very much in that understated "Early-Landed" style.

It works rather well as a Ti Punch, with the lime adding to the citric elements and the sweetness just cutting back on the funk.

TASTING VALUES		
4	Ti Punch	

RHUM JM 2003 (BOTTLED 2014) 44.8% ABV

The JM Distillery dates to 1845, when Jean-Marie Martin bought the Fonds Préville sugar plantation in the far northwest of Martinique. It makes some of the most singular of the *agricoles*, and for me, the ones that appear to need the most time to develop fully.

The fruits typical of the distillery, especially pineapple, come though more strongly here than on the *blanc*, which helps to counteract the tannin. The *rhum* has also become more perfumed after the extra time in the cask, although that white-pepper heart remains, as do some touches of dry grasses. It, like Neisson, is a fine example of terroir at work.

The palate has real elegance and development, with some clove touches, light citrus oil hints, and an oozing quality that still grips and fill out on the mid-palate, giving it precision, power, and length.

As a Ti Punch, things are more forward but also slightly drier, with more of the oak and spices like green cardamom that were hinted at when neat coming through. Stirred down a little more and things become more harmonized, refined, and spicy before a sweet kiss on the end.

TASTING VALUES		
4	Ti Punch	

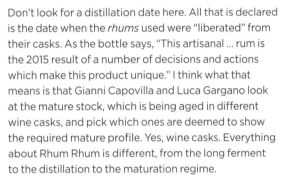

RHUM RHUM LIBERATION 2015 58.4% ABV

Don't look for a distillation date here. All that is declared is the date when the *rhums* used were "liberated" from their casks. As the bottle says, "This artisanal ... rum is the 2015 result of a number of decisions and actions which make this product unique." I think what that means is that Gianni Capovilla and Luca Gargano look at the mature stock, which is being aged in different wine casks, and pick which ones are deemed to show the required mature profile. Yes, wine casks. Everything about Rhum Rhum is different, from the long ferment to the distillation to the maturation regime.

It has a big and elegant nose, with some beeswax, leather – suggestive of old mature stock – as well as Brazil nut tones and a little oiliness. This then moves to muguet, cigar box, lightly smoky oak, and a little peach on the back palate before it begins to dry, picking up the heady, dusty slightly funky perfume of the African bush. The palate also has this slightly burned quality alongside tobacco, apple, sumac, pomegranate, and some acidity. A big, wild, and untamed spirit that expresses its place.

As a Ti Punch it is pretty singed, with some heavy toast and char from the wood. The lime is in check, but does help to add some top notes to what is pretty leathery and resinous. It's probably best on its own.

TASTING VALUES		
N/A	Ti Punch	

WORLD RUM

Wherever cane grows you will find rum, and wherever rum is made you will find variations on that simple theme. That means that this section is a bit of a catch-all, with little commonality between the rums on show. It does, however, show the international nature of the spirit. It is also a chance to have a look at blending: taking disparate styles and melding them into a coherent and complex whole, or using different ageing techniques and climatic conditions to create new flavours. It shows how the spirit of innovation and the drive for premium aren't restricted to the Caribbean but are found around the world.

Finally, it reminds us that when it comes to rum-drinking, India leads the way. Rum is the subcontinent's native spirit, and some of the world's biggest-selling rum brands are made there. Welcome to the world.

AMRUT OLD PORT 40% ABV

Bangalore's Amrut Distilleries (*amrut* means "nectar of the gods") was founded in 1947 as Amrut Laboratories by JN Radhakrishna. The firm began to distil in 1948. Today is it probably best known internationally for the quality of its single-malt whiskies, but its Old Port rum brand is currently the eighth largest-selling rum in the world.

The spirit of innovation that has driven the whisky side of the business is also evident in Amrut's rum division. Two Indies Rum – a blend of rums from Jamaica, Barbados, and Guyana with jaggery-based Indian rum – is well worth seeking out. Old Port is a molasses-based, column-still rum. The nose opens with rich aromatics, mixing blackcurrant jam with orange marmalade, molasses, a hint of sweet nut, black banana, and some cassia. The palate is quite light, with this fragrant note continuing: there's Turkish Delight, chocolate bitters, some ginger, and lime chocolate. Though medium-sweet, there's a touch of tannic grip.

Coconut water is strangely cheesy, with the coconut lying on top. Cola opens nicely, but fades; I'd have it short and hard. Ginger beer is a little oily but offers length and some chocolate, while clementine juice is fresh, rounded, and quite unctuous.

As an Old-Fashioned, things go back to the marmalade, with added hot tiles and a bit of funk. It's slightly sweet, with some Juicy Fruit gum flavours. Chew on that.

TASTING VALUES			
2	Coconut Water	3.5	Clementine Juice
3.5	Ginger Beer	3	Cola
3	Old-Fashioned		

BANKS 7 GOLDEN AGE BLEND 43% ABV

Named after botanist Sir Joseph Banks, this brand was created in 2008. As far as I can ascertain, Banks wasn't a noted rum-drinker, but he did wander the world discovering its natural marvels, and you could argue that the Banks team has done much the same with its rums.

This is a blend of 23 rums from eight distilleries and seven countries: Trinidad, Jamaican pot still, Guyana, Barbados, Guatemala, Panama, and some Batavian arrack (the "seven" on the label refers to the number of countries). The nose has fruit, pot-still funk, and a nutty start before moving into honey, hibiscus, mace, and cinnamon. All the elements are balanced and working for the benefit of the whole. The palate is fat, rounded, and sweet to start, with good pot-still weight, a concentration of fruits, and a sweetness that is always balanced by oak. Multifaceted and supple, it's an exemplary blend.

The slight issue where mixers are concerned is with cola, which is dominant. Ginger beer shows baked fruits and a release of fresh ginger on the end. Clementine juice links itself on the mid-palate, but adds a long, creamy, tropical finish. Coconut water brings out the arrack and pot-still mix, coconut flesh, and a connection to the nuts and oak. The Old-Fashioned is elegant, with drier tones working with the bitters, the sugar adding some softness. It's expansive and complex – and highly recommended.

TASTING VALUES			
4	Coconut Water	4	Clementine Juice
4	Ginger Beer	3	Cola
5*	Old-Fashioned		

BUNDABERG SMALL BATCH
40% ABV

It's only when you go to the Bundaberg Distillery in Queensland that you begin to comprehend the obsessive love its fans have for the brand. No other rum engenders such insane levels of fervour or has quite the same style of hogo. It has been made since 1888, using molasses made from the local cane fields, and while Bundy could never be accused of taking itself seriously, in recent years it has made an effort to move beyond the bogan image and establish a premium range.

The molasses are given a 36-hour ferment. The wash is then distilled in a single column, and either redistilled in one of three pots (with cast-iron bases) to make "heavy", or run through a copper rectifying column, giving light. It's that pot still that gives that Australian hogo. Small Batch is a blend of the two distillates aged in ex-Cognac and Australian brandy casks. It has a rich molasses note, with dried orchard fruit, sweet spices, menthol, and, with water, dried flowers. This potpourri element continues on the palate, alongside dark chocolate, a funky midsection, then a fruity, spiced finish.

With cola things get too muddled; coconut water is molasses-rich, but too hefty; ginger beer is sassy and vibrant, while clementine juice is voluptuous and filled with tinned peach flavours. In an Old-Fashioned there's a gear shift into citrus oils, making for a clean, balanced mix. This is a new range to watch.

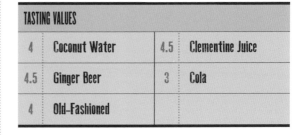

TASTING VALUES				
4	Coconut Water	4.5	Clementine Juice
4.5	Ginger Beer	3	Cola
4	Old-Fashioned		

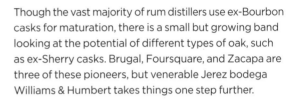

DOS MADERAS PX 5+5
40% ABV

Though the vast majority of rum distillers use ex-Bourbon casks for maturation, there is a small but growing band looking at the potential of different types of oak, such as ex-Sherry casks. Brugal, Foursquare, and Zacapa are three of these pioneers, but venerable Jerez bodega Williams & Humbert takes things one step further.

This expression of its Dos Maderas range is a blend of Bajan and Guyanese rums, aged separately for five years before being shipped to Jerez, Spain, where they're blended and aged in casks that had previously held Dos Cortados Palo Cortado Sherry. After three years, the rum is then given two more years in ex-Don Guido PX (Pedro Ximénez) butts. The nose has the big, dried fruit quality of PX, as well as tamarind, resin, and an oxidized nuttiness, then big Demerara-style notes. The Sherry adds acidity, sweetness, and structure. The palate is rich and heavy, with a concentrated raisin element and some grip. Savoury, yet sweet, this is more of a dessert rum than an afternoon one.

Cola rolls over and dies, simply amplifying the PX. Ginger beer shows some texture and drive, but ultimately founders, as does clementine juice: raisin and juice aren't happy partners. And while coconut water has a certain brio, again the PX acts as a barrier. The Old-Fashioned should be made with chocolate bitters, but to be honest this is best on its own, with a little ice. Lovely rum.

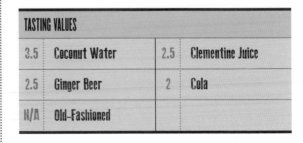

TASTING VALUES			
3.5	Coconut Water	2.5	Clementine Juice
2.5	Ginger Beer	2	Cola
N/A	Old-Fashioned		

DOS MADERAS, LUXUS DOBLE CRIANZA 40% ABV

For this super-premium expression of its range, Williams & Humbert has taken 10-year-old Demerara and Bajan rums, both aged separately in the Caribbean, then shipped them to Jerez where they are given five years in butts that previously held Don Guido PX, a Sherry made from dried Pedro Ximénez grapes. The Sherry is given a very slow movement through a solera system, building up its power and resonance.

The nose is massive, all black cherry, peach stone, and marzipan behind which is the slightly oily weight of the rum that itself has heft, balance, and a powerful elegance. Complex, heady, and aromatic, it bursts on the tongue with tones of citrus, raisins, then star anise and sloe, which brings to mind the Basque speciality Pacharán. This wild cherry/sloe element continues on the thick, dense, and mouth-coating palate, adding liquorice, mulberry, and treacle flavours. There's less grip than the 5+5 (*see* opposite).

It is a singular rum, an after-dinner rum. It is not, on this evidence, one to be mixed with – at all. It is, however, one to be tried. The firm says that it's a rum for "the most sybaritic", and it's a fair description. Only hedonists need apply here. Form a queue!

TASTING VALUES			
N/A	Coconut Water	N/A	Clementine Juice
N/A	Ginger Beer	N/A	Cola
N/A	Old-Fashioned		

MCDOWELL'S NO 1 CELEBRATION 42.8% ABV

Angus McDowell was a Scottish trader who established a company specializing in alcohol and cigars in Madras (now Chennai) in 1826. It was bought by Vittal Mallya's United Breweries (UB) in 1951. In 1959, UB began to diversify into the spirits business, building distilleries across the subcontinent and launching brands, including three under the McDowell's No 1 label, starting with a brandy (1963), then a whisky (1968). McDowell's No 1 rum appeared in 1990 in two styles: Caribbean (white rum) and Celebration (dark). It is currently the world's largest-selling rum, at 19 million cases a year and counting.

The Celebration has a nose of fresh molasses with a slight iron/blood tang. Firm and dry, aromas suggest burned toast, well-fired hot cross buns, then dried prune. The palate is thick and sugary to start, with notes of cola, vanilla, then a green distillate edge.

Maybe it's because it already smells of cola, but this mixer adds a hollow quality. Ginger beer is little better – more akin to ginger wine – but is drying. Clementine juice has an overripe-fruit quality, yet remains dumb and short. Coconut water works by steering the rum in a new direction. There's still masses of treacle, but the mixer adds length and balance. An Old-Fashioned is all Camp coffee sweetened with brown sugar.

TASTING VALUES			
4	Coconut Water	2.5	Clementine Juice
2.5	Ginger Beer	2	Cola
2.5	Old-Fashioned		

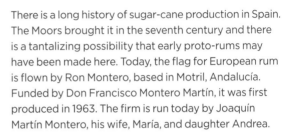

RON MONTERO
GRAN RESERVA 40% ABV

There is a long history of sugar-cane production in Spain. The Moors brought it in the seventh century and there is a tantalizing possibility that early proto-rums may have been made here. Today, the flag for European rum is flown by Ron Montero, based in Motril, Andalucía. Funded by Don Francisco Montero Martín, it was first produced in 1963. The firm is run today by Joaquín Martín Montero, his wife, María, and daughter Andrea.

Using imported molasses, two distillates are made in a four-column system: one at 80% ABV the other at 96% ABV. These are aged separately in soleras of American oak butts before being blended and bottled. The rum opens with a smoky edge, then come sweet spices (cumin seed), honey, and a fresh, green element – olive, cane leaf, and dry grass – and sweet concentration. The palate is clean and delicate, mixing the natural spiciness and oak well before lemon thyme gives a final lift.

Coconut water becomes almost too green and leafy, with a separation of sweet and dry, with the latter dominating. With cola the rum comes into its own, with sweet richness on show. Ginger beer is pure ginger juice, with rum in a supporting role, while clementine juice is a cheerful mix. An Old-Fashioned brings out more exotic spice, the rum showing almost Cuban-style dryness, with good drive on the finish. A fine sharpener.

TASTING VALUES			
2	Coconut Water	3.5	Clementine Juice
3	Ginger Beer	3	Cola
4	Old-Fashioned		

OLD MONK 7-YEAR-OLD
40% ABV

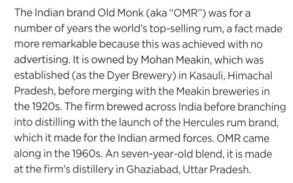

The Indian brand Old Monk (aka "OMR") was for a number of years the world's top-selling rum, a fact made more remarkable because this was achieved with no advertising. It is owned by Mohan Meakin, which was established (as the Dyer Brewery) in Kasauli, Himachal Pradesh, before merging with the Meakin breweries in the 1920s. The firm brewed across India before branching into distilling with the launch of the Hercules rum brand, which it made for the Indian armed forces. OMR came along in the 1960s. An seven-year-old blend, it is made at the firm's distillery in Ghaziabad, Uttar Pradesh.

The nose is hefty and molasses-rich, with notes of treacle scone, prune juice, and bitter chocolate. With water things become dusty and more like a traditional cough medicine. The palate is a mix of clove oil, raisin, sugar, and rich spirit. It's in the military style: violet chocolate with sweetness and bitterness playing off each other.

Coconut water is a step too far, with the mixer devoured by the treacle. Ginger beer is calmer and a decent match, while cola is a pairing of equals, so if you like your drinks sweet... Clementine juice brings in overripe fruits and syrups.

Mixing into an Old-Fashioned makes this taste like a Rum and Black (the Navy rum and blackcurrant cordial drink beloved of old sailors in Britain).

TASTING VALUES			
2.5	Coconut Water	3.5	Clementine Juice
3	Ginger Beer	3.5	Cola
2.5	Old-Fashioned		

THE RUMS: WORLD RUM

PENNY BLUE XO SINGLE ESTATE BATCH 004 43.3% ABV

Rum may be the last thing you would expect a renowned London wine merchant established in 1698 to be making, but Berry Bros & Rudd has always cast its net wider than most of its associates. It owned Cutty Sark, makes No 3 gin, and is a respected independent bottler of single-malt Scotch. The firm's spirits "nose", Doug McIvor, is behind this rum, along with Jean-François Koenig, master distiller at the Medine Distillery in Mauritius.

A blend of three distillates – molasses, cane syrup, and cane juice – fermented in two stages for 48 hours, distilled in a four-column continuous still to 95% ABV, it is aged in ex-Cognac, -Bourbon, and -Scotch casks. The rums are either blended or sold as single casks. Each batch is slightly different. The nose opens with a mix of lightly oxidized Oolong tea, red berries, floral notes, and citrus peels before developing quince and lemongrass notes. The palate is gentle and direct, with white pepper and a hint of cane opening well on the back palate, which shows concentration, boiled sweets, mint, and spice.

Cola is a little blunt; ginger beer has a central energy but not enough heft; but coconut water is refined and subtle with good length, while clementine juice opens up the rum's fruitier side. An Old-Fashioned is almost barley sugar with clove, the rum enhanced and making a calm and gentle mix. Recommended.

TASTING VALUES			
4.5	Coconut Water	5	Clementine Juice
3.5	Ginger Beer	3	Cola
4.5	Old-Fashioned		

THE RUMS: WORLD RUM

NAVY & DARK RUM

If there is one term guaranteed to get most rum distillers irate, it is calling their aged rums "dark". It is a term that has become pejorative, a throwback to the Bad Old Days of rum, when the only alternative to white was a rum style modelled on the nineteenth-century rums drunk by the British Navy. "We have moved on," they protest. "These are just coloured-up rums pretending to be old. They are nothing like what we produce." Well, yes. They have a point, but that shouldn't mean that we then regard this style of rum as lacking in validity.

These rums are darker in hue: reddish rather than "dark". Molasses and caramel are more apparent. Black fruits are everywhere. There's often a bittersweet edge. The base spirit is young. That is what they are. Deal with it, because the best examples here have balance, have character, and are versatile.

They are also quirky. No mixer here outperformed any other. For every rum that blossomed with ginger beer, there was another that spurned it. The moral? Judge these rums on an individual basis.

BLACKWELL BLACK GOLD
40% ABV

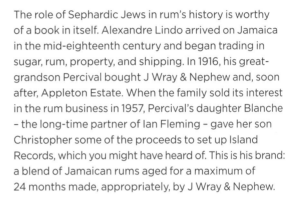

The role of Sephardic Jews in rum's history is worthy of a book in itself. Alexandre Lindo arrived on Jamaica in the mid-eighteenth century and began trading in sugar, rum, property, and shipping. In 1916, his great-grandson Percival bought J Wray & Nephew and, soon after, Appleton Estate. When the family sold its interest in the rum business in 1957, Percival's daughter Blanche – the long-time partner of Ian Fleming – gave her son Christopher some of the proceeds to set up Island Records, which you might have heard of. This is his brand: a blend of Jamaican rums aged for a maximum of 24 months made, appropriately, by J Wray & Nephew.

A treacly nose opens into strawberry, chocolate, and some walnut tones, with solid Jamaican depth, here going into black banana and raisin. The palate is thick, with coffee grounds and an almost earthy depth alongside squished black fruits to add sweetness. There's good grip, suggesting that the oak's already having a say.

Rich, with a certain swagger, this comes to life when mixed. Coconut water is the star, creating a stunning hedonistic drink with sweet, velvety depths. Ginger beer is punchier, but equally long, while cola benefits from the molasses to make a drink with sweetness, fruit, and depth. While clementine juice might not reach the same heights (or depths), it's a muscular mix. Keep it simple is the key here. An Old-Fashioned is all overcooked damson jam.

TASTING VALUES			
5*	Coconut Water	4	Clementine Juice
5	Ginger Beer	5	Cola
3	Old-Fashioned		

GOSLING'S BLACK SEAL BERMUDA BLACK RUM 40% ABV

The Gosling brothers, James and Ambrose, opened a shop in Hamilton, Bermuda in 1824. Thirty-four years later they started to blend rum, bottling it in recycled Champagne bottles from the Navy's officers' mess. Today, it is a blend of 97 per cent column-still rum and three per cent pot-still rum, both aged for a minimum of three years. The Navy started mixing Gosling brothers' rum with ginger beer when the Admiralty opened a ginger beer bottling plant at Ireland Island Dockyard in Bermuda. Today, the Dark 'n' Stormy is a Gosling's registered trademark.

The nose opens with roasted notes, a light nuttiness, lime hulls, and mentholated balsam. The palate is sweet and slightly confected, with red and black fruits, cherry pie, and molasses. With water, there are throat lozenges, and some crystallized fruits.

There's life beyond the default drink. If the clementine juice is a little stewed and its acidity lost, cola has a bittersweet depth that adds character. With coconut water, the molasses acts as an anchor, the salty notes adding a new dimension. Still, it's the Dark 'n' Stormy that rules. Peppy and deep, it has a liquorice, treacly length and a tingling long finish. It's a tougher ask as an Old-Fashioned, even if the orange adds life. Quite fat and slightly bittersweet, it's neither diminished nor enhanced.

TASTING VALUES				
4	Coconut Water	3.5	Clementine Juice
5	Ginger Beer	4	Cola
3	Old-Fashioned		

LAMB'S NAVY RUM 40% ABV

By the nineteenth century, London's docklands were full of rum merchants all making their own signature blends, most based around the style of rum popularized by the Navy, which, by the middle of the century, was beginning to give greater prominence to the Demerara style.

Take Alfred Lamb's blend as an example. First made in 1849 as a blend of rums from British Guyana, Barbados, Trinidad, and Jamaica, it is now 100 per cent Guyana; to be precise, it is a two-year-old blend of Savalle multiple-column and Port Mourant double wooden pot rum.

The nose is a mix of heavy treacle, black banana, roasted almond, and cacao. Dark in colour, it's actually quite light in nature and becomes burned and lightly phenolic with water. On the palate, you pick up rhubarb, Christmas cake, and Porter beer. Quite fresh with water, there's some Danish liquorice and light tannins. This is a solid example of its style.

As a British-style rum, it's maybe no great surprise that coconut water isn't up to much, and while clementine juice makes it fruity, the palate is dry and bitter. Cola makes things chewier; its fruit and leather tones create a decent, if straightforward drink. It gives ginger beer a slightly mysterious air with, for the only time, length, as that bitter edge works well with the spice.

It's probably best to draw a veil over the Old-Fashioned – unless you like burned Christmas puddings.

TASTING VALUES			
2.5	Coconut Water	3	Clementine Juice
4	Ginger Beer	3.5	Cola
2	Old-Fashioned		

MYERS'S ORIGINAL DARK
40% ABV

Because Isaac Myers of Portsea was a supplier of goods to the British Navy, it's perhaps no surprise that when his sons, Michael S and Fred L, moved to Jamaica at the start of the nineteenth century they began trading in rum – thereby establishing another link between the spirit and the Jewish community. Fred L Myers launched his own brand in 1879, the portfolio being bought by Canadian giant Seagram in the 1950s. It is now part of Diageo.

The Jamaican roots remain strong. Myers's blends nine molasses-based pot and column-still rums of up to four years of age and is best known as the base for the Planter's Punch cocktail that appeared in the 1920s. Dense, and slightly bitter with an almost burned aroma – think grilled red pepper and molasses – there's some uplift into wet raffia, chamois leather, and a touch of Jamaican funkiness, which then begins to control proceedings. It's the rummiest rum many will have encountered. The palate mixes molasses with grassiness, some acidity, and a thump of concentrated fruit. It's also dry on the end, with some pot-still overtones.

You know where you are with Myers's. If ginger beer proves to be strangely bitter, all the other mixers are fine: slightly funky in coconut water's case, marmalade-like in clementine juice's, and with coffee accents and some density with cola. An Old-Fashioned, however, needs more complexity to work.

TASTING VALUES			
3	Coconut Water	3	Clementine Juice
3.5	Ginger Beer	3.5	Cola
2.5	Old-Fashioned		

OVD OLD VATTED DEMERARA
40% ABV

George Morton Ltd was a Dundee-based firm located in the city's Dock Street, and later in the Chapel Works buildings in Montrose. Like many merchants of the Victorian era, George Morton began by experimenting with blending, and while the firm was to become known for its Scotches, it was rum that started it all. His OVD dates back to 1838 – 13 years earlier than the first blended Scotch brand. The letters stand for Old Vatted Demerara (the firm also had a Jamaican-based blend called OVJ).

Opening with a typical liquorice/molasses aroma, this has some oak on show from the outset, with notes of chicory coffee, old forest, turned earth, then minty chocolate and cocoa nibs. This isn't a heavy rum, but it does deliver a solid and reassuring thunk of molasses halfway through. The finish hints at a burned-sugar note and some red fruit. A well-made example of this old style of rum.

For once, though, cola doesn't work – it makes the mix too sweet and unbalanced. Coconut water creates a slightly autumnal, but clean and balanced cocktail. While the fruits come through with the clementine juice, things are a little tight in the middle, yet it's a decent mix. Ginger beer is all root-ginger freshness with some weight in the centre, and it's the one I'd reach for. An Old-Fashioned ends up down the local curry house, but there's not enough weight to add any complexity to this tricky mix.

TASTING VALUES			
3.5	Coconut Water	3.5	Clementine Juice
4	Ginger Beer	2.5	Cola
3	Old-Fashioned		

PUSSER'S GUNPOWDER PROOF RUM 54.5% ABV

On July 31, 1970, when the final tot of the British Naval rum ration was poured, a very specific style of rum ended. Yet no one had counted on the perseverance of Charles Tobias, who took nine years to persuade the Admiralty to give him the rights and the recipe for making Naval rum. He called the new brand Pusser's, in honour of the purser who oversaw the serving of the daily tot.

A pot-still-rich blend of rums from Guyana and Trinidad, aged for up to three years, this Gunpowder Proof is at the same strength as would have been dispensed aboard ship. It has concentrated currant fruits, with sloe and chocolate ganache tones rounded off with nutmeg, cream, and caramel syrup. There's even more pot-still weight when diluted, creating an explosive palate full of fruit pastilles and rolling Demerara mellowness. Though sweet, it has sufficient grunt to balance. With water, the heavier elements coalesce in the middle, allowing delicate citric and tropical notes to free themselves.

Heavier rums can struggle with coconut water, and while there's some layering here, the tannins get in the way. Apricot comes out in the clementine juice mix, as well as a bitter edge. Ginger beer adds real effervescence, while cola becomes all deep-earth tones, making for a rather fine drink. The Old-Fashioned is thick with chocolate and char, but ends up trailing this burned edge.

TASTING VALUES			
2.5	Coconut Water	3	Clementine Juice
4	Ginger Beer	4.5	Cola
2.5	Old-Fashioned		

CACHAÇA (& AN OUTLIER)

Would that there were space to include more cachaça in the book. Would that there were more cachaças available outside of Brazil to be able to try. It is a great irony that one of the world's most widely consumed spirits is still considered a "speciality". It has also been cachaça's fate that, until recently, the brands that have been most widely available on the export markets have been the large industrial brands, whose quality (or lack of it) has not done anything to endear the spirit to an interested drinker.

Things are changing. The major international drinks firms are buying up cachaça brands, and some artisanal brands are beginning to appear. This can only be a good thing, because at its best, cachaça is a fascinating and complex spirit which can be enjoyed mixed or on its own. The brands that follow are therefore just an amuse-bouche for what, if you have any sense (and I know you do), will become a minor obsession. Here, I stuck to the principle that white cachaça is served in a Caipirinha, while the aged examples are enjoyed on their own. Don't take this as a rule, however; feel free to play.

This section also seems to be the best place to slot in Batavia-Arrack van Oosten, the cane-based revenant that drives bartenders wild.

ABELHA SILVER 39% ABV

Owned by the Responsible Trading Company, this organic cachaça uses cane from the sandy soils of Chapada Diamantina National Park in Bahia. It is fermented with a sugar-cane yeast cultured from the farm's own cane. It is then distilled in a small, 400-litre (333 gallon), copper-pot still before being rested in open stainless-steel tanks for six months to allow the more aggressive compounds to flash off. It's then reduced in strength with water from an aquifer under the farm.

The nose has fresh cane notes from the off that hints at *agricole*, but has more of a green-fig element alongside lychee, and some lily tones. When reduced, a brioche/pastry note emerges. This is fruit-dominant, with a clean lime acidity running through it.

The palate is soft and well balanced, with a fleshy texture to the mid-palate that dries slightly on the end. The juiciness at its core comes through at the back of the nose and with water, where the flowers are dialed down and the fruit elements increased.

You still get a good punch of spirit in the Caipirinha; personally, I like it, as it adds character. This has weight, length, balance, and persistence. Recommended.

TASTING VALUES		
5	Caipirinha	

LEBLON 40% ABV

The fact that major international drinks firms are snapping up cachaça houses speaks volumes of their belief that this should become a global spirit. The most recent (at time of writing) has been Bacardi's purchase of Leblon. This brand was established in 2005 by Steve Luttman, former senior vice-president of marketing at LVMH, and his Brazilian father-in-law. Their aim? To educate people that cachaça wasn't industrial firewater. Luttman promptly hired Gilles Merlet from Cognac as his master distiller and got to work.

Leblon is an estate-grown artisanal cachaça, made using hand-cut cane grown in the firm's estate in Patos de Minas in Minas Gerais. Distilled in pot stills, it is given six months resting in ex-Cognac casks.

The nose is calm and almost serene, with good cane notes, a little tropical hit, and peanutty underpinnings (conceivably from the oak). It then moves cleanly into pear, mixing acidity and softness. The palate is light, and surprisingly dry when neat, though softer when water is added.

In a Caipirinha, it gains extra lift but retains its inherent delicacy, while the more vegetal notes begin to emerge. The sugar helps to fill out the mid-palate and there's a slightly sappy, fresh finish.

TASTING VALUES		
4.5	Caipirinha	

SAGATIBA PURA 38% ABV

It is fair to say that what Sagatiba has done for people's attitude to cachaça is similar to what Patrón did for new drinkers' notions about tequila, or how Bombay Sapphire brought people into gin. It's the gateway cachaça.

Now owned by Campari, the brand was the brainchild of Marcos de Moraes, who saw the opportunity for a new kind of cachaça: lighter in style, premium, less "challenging" than the big industrial brands, but with more volume than the artisanal examples.

It is a column-still distillate, which is redistilled in a second column with various reflux plates, allowing specific flavours to be removed. It is taken off at bottling strength rather than being reduced. No sugar is added.

The nose is fresh and green, with touches of sourdough sponge; even at 38% ABV it has a clean punch, with a little hint of geranium leaf and very light cane juice. The palate has the bland quality of ripe pear cut with zippy acidity and a slightly peppery edge. It's light, clean, and well balanced.

It makes a zesty and fresh Caipirinha, with a rare mineral quality. It's light certainly, but it's also refreshing.

TASTING VALUES		
4	Caipirinha	

ABELHA ORGANIC 3-YEAR-OLD 38% ABV

This aged example from the organic cane fields of Bahia has spent three years in 250-litre (55 gallon) garapeira (*Apuleia leiocarpa*) casks. Otherwise known as "yellow wood", garapeira is widespread in Bahia, where it is widely used in construction. It is said to impart a honeyed element to the spirit.

The nose is in line with good intensity, but more deep, super-ripe fruits are on show, oozing sugars. This is very clean, with a touch of tomato from the cane juice. The palate is light- to medium-bodied, with fresh, green acidity and butter-fried hot spice. It manages to be both bold and relaxed in its aspect, with all its energy concentrated in the centre.

Water brings out a nuttier and, yes, honeyed element, with a fruit-skin bloom of yeastiness on the back. Complex and very good.

MAGNÍFICA, LAS IGUANAS
37% ABV

This aged cachaça is made for the UK restaurant chain Las Iguanas by João Luiz Coutinho de Faria at the distillery located on his farm, Fazenda do Anil, on the border between Vassouras and Miguel Pereira, close(ish) to Rio de Janeiro. It uses hand-harvested, estate-grown cane and is distilled in one of only two three-chamber pot stills in Brazil (known as *alegría*).

These operate in a similar fashion to a pot-and-retort system, with three linked pots at different levels. All are filled with "wine". The lowest pot is heated, driving off the alcohol vapours into the middle pot, where the process is repeated. The now-enriched vapour stream then rises to the top pot, where it is carried in a pipe to a condensing worm that diverts it to the full condensing system. The heat from this vapour then warms the "wine" in the top pot.

The colour is pale green and has more apparent oak notes than some aged cachaças. The aromas bring to mind wet mackintoshes, grapeseed, some moss, and the humid smell of an orchid house – fresh, aromatic, and complex. The palate has a big, waxy, cocoa butter start and a gentle flow with a little funky depth. In time, and with water, it moves into mint and fresh oregano, with a more relaxed air to it than when neat. Forget this idea of cachaça being harsh.

YPIÓCA OURO  39% ABV

You couldn't be further away from the artisanal world with Ypióca, which is made by Brazil's largest producer, based in Cereá. It has been part of multinational group Diageo since 2012. This example has been given between two and three years ageing in freijo (*Cordia alliodora*) casks. The wood, if you are interested, is often used for making guitars.

It has a light greenish cast and a very soft nose that brings to mind tinned pineapple juice that's cut with a sappy green note, and hints of lambic beer. In turn it becomes slightly floral (stems), then chalky. A bready note develops with water, alongside nettles and cumin.

The palate is light, with dry spices, a little chocolate, some heat, and a touch of grip on the back palate. In time, the chocolate builds in strength before the cane juice flows through, along with a little chestnut mushroom. Fairly hard. Add ice.

BATAVIA-ARRACK VAN OOSTEN 50% ABV

Bartenders wishing to replicate the authentic taste of eighteenth- and nineteenth-century punches had a dilemma. Many of the recipes called for arrack. The problem was, it was seemingly impossible to get hold of – an old spirit that had ceased to exist. Enter on one side importer Eric Seed of Haus Alpenz, and *éminence grise* of all things cocktailian, David Wondrich; on the other, E&A Scheer of Amsterdam, which had been involved in the arrack trade from 1818 and which held stock destined for flavourings and Swedish *punsch*. The result of their joint efforts was this brand.

Produced in Java, arrack is molasses based, with a ferment that is triggered by the use of a "starter" made of red rice cakes. Distillation is in Chinese-style pot stills before the spirit mellows in large teak vats.

The nose is off-dry, with a weird, funky meatiness mixed with cooked vegetable, tofu, tinned borlotti beans, and wet silk before more orthodox hints of chocolate and thyme come through, followed by molasses.

The palate is thick with a yeasty, oxidized, *flor*-like flavour that brings to mind *vin jaune*. The funk emerges in the slightly sweet mid-palate, which combines genmaicha (Japanese green tea mixed with brown rice), black olive, and an prickly, spicy energy that gives some heat on the finish. The feel is as important as the aroma here.

This is an essential addition to any serious rum-lover's spirits cabinet.

COCKTAILS

Is there any other spirit that so eagerly flings itself into the world of mixed drinks? Rum punch dates back to the 1600s; British sailors were drinking proto-Daiquiris the century after. It was rum that filled the shakers and blenders of Cuba's innovative *cantineros* from the 1920s onwards and, most obviously, gave America a sugar-laden boost of happiness in the First Age of Tiki.

What follows are recipes old and new from long-gone classic bartenders and a selection from some of the world's top contemporary shakers. All are doable at home, as long as you remember these age-old rules:

- Don't get yourself dragged into making a selection of different drinks at a house party. Make a punch instead.
- Keep things simple.
- Know your rum and let it sing. These are rum drinks, not fruit drinks.
- Understand balance. The recipes that follow might need a tweak or two to suit the flavours of the rum you are using, your ice, and the other ingredients.

Above all, have fun.

RECIPES

Punch-making isn't about randomly throwing together booze and seeing what happens. If you do that, people will start reeling after two glasses, then fall over. Punches are sociable, meant to be shared, and are diluted, which makes them balanced, easy-to-drink libations. The rum needs to have weight, and that means using arrack, pot still, or *agricole*.

Punches have the funk, and the funk is its own reward.

O'DOHERTY'S ARRACK PUNCH

Makes 3 cups

60ml (2fl oz) demerara sugar

60ml (2fl oz) boiling water

60ml (2fl oz) lime juice

180ml (6fl oz) Batavia-Arrack van Oosten

120ml (4fl oz) Smith & Cross Traditional Jamaica Rum

360ml (12fl oz) cold water

cubed ice

grated nutmeg, to serve

In a large pitcher, dissolve the sugar in the boiling water. Add the lime juice and stir. Add the arrack, rum, and cold water. Refrigerate until ready to use. Serve in a punchbowl with ice cubes. Grate nutmeg on each cup as you serve.

Adapted from an 1824 recipe by David Wondrich in *Punch*.

RUM PUNCHES

Come all you bold heroes, give an ear to my song
And I'll sing in the praise of good brandy and rum
There's a clear crystal fountain near England shall roll
Give me the punch ladle, I'll fathom the bowl.
Traditional English folk song

Our mixed-drink tale starts with punch, which itself originates in the seventeenth century with thirsty traders of the English East India Company in India and their equivalents in the Dutch East India Company (aka the VOC) further east mixing spirit – maybe imported, but often locally distilled arrack from palm or cane – with a souring agent, sweetener, diluting agent (water and/or tea), and spice. These pillars continue to support the temple of punch. Get the balance right between them all and you have a magnificent libation.

As there is no evidence of widespread punch-drinking among the native population at the time records begin, it was either a mix that spontaneously appeared, or was one that arrived by other means. All fingers point to it being waterborne or -born. The British Navy had been carrying spirit on board ships since the sixteenth century to cure ills, but also to carouse with. In fact, all the ingredients for punch would have been on board.

It's therefore probably the Navy that brought punch-drinking to the new plantations in the Caribbean, where by the start of the eighteenth century, according to physician Sir Hans Sloane, punch had become "the common fuddling liquor of the more ordinary sort". Its low-class image was not to last, however. The eighteenth century was a time of "bowl-fathoming" as a direct consequence of rum (and arrack's) rise in status. In Britain, rum punch was a drink for coffee houses, gentlemen's clubs, country houses, and taverns. It was both fashionable and somewhat decadent, reeking of the funk of wealth and the Indies. It also elevated you from the masses supping their gin. If you drank punch, you had time to kill – stories to tell.

COCKTAILS

GLASGOW PUNCH

This is a classic eighteenth-century planter-style punch made famous in a city that grew fat on the Atlantic trade, and was famous for its punch – and its drinking. Makes 5 cups

170g (6oz) caster sugar
....
170ml (6fl oz) water
....
120ml (4fl oz) lemon juice
....
570ml (20fl oz) cold water
....
210ml (7fl oz) Jamaican rum
....
2 limes, halved
....

In a bowl, dissolve the sugar in the water. Add the lemon juice and cold water. Stir in the rum. Run the lime segments around the rim of the bowl, then squeeze in the juice.

TIKI BOWL

A shared tiki drink which is basically a scaled-down punch.

30ml (1fl oz) light Puerto Rican rum
....
30ml (1fl oz) dark Jamaican rum
....
30ml (1fl oz) VSOP Cognac
....
60ml (2fl oz) orange juice
....
45ml (1½fl oz) fresh lime juice
....
15ml (½fl oz) orgeat syrup
....
170g (6oz) crushed ice
....

Blend all ingredients together for around 10 seconds. Serve in a tiki bowl with two straws.

From *Trader Vic's Bartender's Guide*, 1972, with thanks to Beachbum Berry.

In the American colonies, rum punch was more egalitarian, as well as more political. Not only did it usher in the concept of identity, it helped smooth voters' decisions. George Washington engaged less in pork-barrel politics than punchbowl politics to butter up his constituents, while Ben Franklin even wrote an ode to it. Its glorious century was soon to end, however.

Punch declined in the nineteenth century as the restrictive trade barriers that had typified mercantilism began to fall, allowing a greater choice of spirits for all. Additionally, single-serve drinks were becoming more popular: the start of what would become the first age of the cocktail. This had been pioneered by James Ashley, whose London Punch-House opened on Ludgate Hill in 1731 and sold, "the finest and best old Arrack, Rum, and French Brandy... made into Punch". Ashley sold his punch in single measures for "fourpence halfpenny" as well as by the bowl.

The shorter, quicker drinks came into their own in the nineteenth century. Punch is a slow drink, one that promotes relaxed conversation as glasses are dipped into the bowl. It's a drink above which pipe smoke wreathes, from whose depths considered debate can rise – or equally roisterous behaviour. In other words, it is not the drink of the busy capitalist. Still, a hundred years of pre-eminence isn't bad for a drink.

Punch may have declined, but it didn't disappear. All of the great cocktail books of the nineteenth century contained a section on punches, and when tourism arrived in the Caribbean from the 1920s onward, variations on the classic Planter's Punch became the archetypical welcome drink. It was classic punches that formed one of the bases for tiki drinks.

The fathomless bowl continues to brim over.

REGENT'S PUNCH

Makes 10 cups

1 lemon and 1 Seville orange, thinly peeled

115g (4oz) caster sugar

570ml (20fl oz) green tea

60ml (2fl oz) Smith & Cross Traditional Jamaica Rum

45ml (1½fl oz) Plantation Pineapple Rum

60ml (2fl oz) Batavia-Arrack van Oosten

240ml (8oz) VSOP Cognac

60ml (2fl oz) maraschino liqueur

2 bottles brut Champagne or English sparkling wine, such as Ridgeview

First make oleo-saccharum from the lemon and orange: peel the fruit, without any white pith (reserving the fruit for juice) and place in a bowl. Add the sugar, muddle, and allow to rest. Next, in a large bowl, mix the oleo-saccharum with the green tea and juice of the lemon and orange. Set aside for an hour at room temperature, then add the rums, arrack, Cognac, and maraschino liqueur. Strain into a punchbowl, then top up with the Champagne or English Sparkling wine.

This was invented for the Prince Regent (later George IV), a man who loved booze and extravagance and built his stately pleasure dome (or campest building in England, depending on your take) in Brighton, the city where I now reside – hence the suggestion of an excellent local sparkling wine.

WEST INDIAN PLANTER'S PUNCH

Makes 6–8 cups

113g (4oz) caster sugar

113g (4oz) guava jelly

240ml (8fl oz) hot green tea

500ml (18fl oz) aged Jamaican rum

650ml (22fl oz) Cognac

120ml (4fl oz) Madeira

500ml (18fl oz) chilled water

120ml (4fl oz) lime juice

cubed ice

grated nutmeg, to serve

Add the sugar, guava jelly and tea to a punchbowl and whisk until they liquefy. Add the other ingredients and allow to cool. Add ice when ready to serve and grate nutmeg on top.

An 1845 recipe printed in *Beachbum Berry's Potions of the Caribbean.*

CRIMEAN CUP À LA MARMORA

Named after Alfonso Ferrero la Marmora, Crimean war hero and prime minister of Italy. Makes 15 cups

2 lemons, peeled and peels and juice retained

1 bottle soda water

85g (3oz) caster sugar

500ml (18fl oz) orgeat syrup

250ml (8 fl oz) Cognac

120ml (4 fl oz) maraschino liqueur

120ml (4 fl oz) Jamaica rum

cubed ice

1 bottle Champagne

Make an oleo-saccharum from the lemon peel (*see* left). Add the lemon juice, then the soda water, and dissolve the sugar. Add the orgeat syrup and whip. Add the liqueur, strain into a punchbowl filled with large ice cubes. Top with Champagne.

From *Jerry Thomas' Bartender's Guide*, 1862.

COCKTAILS

DAIQUIRI #1

60ml (2fl oz) white rum

15ml (½fl oz) hand-squeezed lime juice

1 tsp caster sugar

chipped ice

"Throw" the drink ingredients between two shaker tins/mixing glasses. One is empty, the other has the drink, the ice, and a strainer on the mouth. Starting with both vessels at eye level (or higher), pour the drink into the empty vessel, slowly drawing it downwards. Tip the liquid back into the ice-filled receptacle and repeat four or five times until the drink is chilled, aerated, and smooth. Strain and serve in a cocktail glass.

DAIQUIRI

It slides in like a rapier, the blade so cold and soft it feels like a caress. There's a moment's pause as the mind freezes, then sweetness smiles briefly before the citrus bites. The alcohol is barely noticed as its two assistants play their game of distraction. It may be hot outside, but you have cooled down, falling slowly as this insidious, seductive drink possesses you.

Rum, citrus juice, sugar. Strong, sweet, sour. Who needs further complication when this trio can work such fell magic? They are the ancient foundation stones of the cathedral of rum drinks.

For Cubans in the nineteenth century, Daiquiri didn't refer to a drink. If they had heard of it at all, they'd have known it as a beachside village, and location of a Yanqui-owned iron-ore mine close to Santiago.

Yet they would have known of the Canchánchara, a mix of *aguardiente*, sweetener (honey or molasses), and *limón*, a drink of the fields, of the people.

The tale of how it became "our" Daiquiri revolves in some way around Mr Jennings Cox, the general manager of the Spanish-American Iron Ore Company, which operated the Daiquiri mine. Depending on who you believe, either one night in 1896 Mr Cox decided to make his guests a gin sour but with no gin in the house, switched to rum; or he instructed the bartender at Santiago's Hotel Venus to make him a rum sour.

It was a good drink, stirred over ice, sweetened with brown sugar, using either lemons or limes (*limón* is Spanish for key lime). It was his drink and eventually he, or someone else, was asked to give it a name. He (or they) chose the name of the village (or mine, or beach).

Does any of this matter? In the macrocosmic scheme of things, no. The fact that there are so many conflicting stories points to three things: it was a good drink, everyone wanted a piece of it, and memories become fuzzy when alcohol is involved.

Fact is, a rum sour already existed. Jerry Thomas lists a St Croix rum-based "Santa Cruz Sour" in the 1887

DAIQUIRI #2

1 tsp white sugar

15ml (½fl oz) lime juice

60ml (2fl oz) white rum

1 tsp curaçao

1 tsp orange juice

chipped ice

Dissolve the sugar in the lime juice. Shake all the ingredients together and strain into a cocktail glass.

reprint of his bartending manual, which the eagle-eyed among you will have spotted was before Mr Cox began stirring things up.

The Daiquiri's success also demonstrates a Cuban paradigm. Most of the country's trends – music being a prime example – start in Santiago and move to Havana, where they are either refined (the Havana take) or commercialized and ruined (the Santiago line). In the Daiquiri's case, I'm with the Habaneros.

Once in the capital, bartenders such as the legendary Maragato replaced brown sugar with white, shook the drink, and used *limón* instead of lemon. It would be another bartender, however, who would make the Daiquiri a classic.

In 1914, a young Catalan bartender joined Bar La Florida in Havana, then owned by Narciso Sala Parera and famed for its "thrown" drinks. Four years later, the young man owned the joint and renamed it Floridita. His name was Constante Ribalaigua Vert. Also behind the stick was Narciso's second cousin, Miguel Boadas, who left Havana to return to his family's ancestral Catalan home in 1927. He opened his eponymous bar in 1933, where he continued to throw drinks the Cuban way.

Constante was a theorizer. He understood the role of ice, ratios, and balance and how to bring complexity within simplicity. Thanks to him, Floridita is known as the *cuna de la Daiquiri* (cradle of the Daiquiri). It is also one of the cradles of classical bartending. He was an artist who also left nothing to chance. Not satisfied with making the perfect Daiquiri, he made five variants, using different techniques, different ice, different tweaks.

I remember watching Alejandro Bolívar, head bartender at Floridita, possibly the man who has made more Daiquiris than anyone else alive, giving a class on Daiquiri #4. "Feed it," he said, liberally pouring rum into the revolving mix. "Don't be scared. Now, listen to the blender. When the tone changes, it's hungry again. Feed it some more." Bartending by sound, intuitive understanding, skill. Subtle complexity. That's the Daiquiri.

COCKTAILS

DAIQUIRI #3

1 tbsp caster sugar

15ml (½fl oz) lime juice

60ml (2fl oz) white rum

1 tsp maraschino liqueur

1 tsp grapefruit juice

340g (12 oz) crushed ice

Dissolve the sugar in the lime juice. Shake all the ingredients together over crushed ice.

Strain into a glass filled with crushed ice. (You could also blend, à la Daiquiri #4.)

VARIATIONS

Ernest Hemingway was a diabetic, so didn't want sugar in his drinks. He was also alcoholic, which meant he wanted booze. The two drinks Constante made for him do just that. The Hemingway is Daiquiri #4 with twice the rum, minus any sweetening. The Papa Doble is a #3, again rum-heavy with sugar eliminated. They're not great drinks – proving that the customer isn't always right.

Fruit Daiquiris are often no more than cheap rum, wet ice, fruit, and a slushy machine, but use good rum, lots of fresh fruit, fresh juice, and good ice and you can make something great. No blender to hand? Muddle, shake, and fine-strain. The mango Daiquiris using fruit just off the tree knocked up by Naren Young and Scotty Schuder after a fishing trip off the Cuban coast will live with me forever. The tamarind Daiquiri in Havana's La Cocinero is a thing of genius.

DAIQUIRI #4

60ml (2fl oz) white rum

15ml (½fl oz) lime juice

5 drops maraschino liqueur

1 tsp caster sugar

340g (12 oz) crushed ice

Blend all the ingredients together with the ice. Serve, unstrained, in a cocktail glass. This is the style most commonly made in the Floridita.

DAIQUIRI #5

60ml (2fl oz) white rum

15ml (½fl oz) lime juice

1 tsp maraschino liqueur

1 tsp grenadine

1 tsp caster sugar

crushed ice

Blend all the ingredients together with ice. Serve, unstrained, in a cocktail glass.

TRADER VIC'S ORIGINAL MAI TAI

60ml (2fl oz) Wray & Nephew 17-Year-Old*

15ml (½fl oz) curaçao

15ml (½fl oz) orgeat syrup

15ml (½fl oz) 2:1 simple syrup (see p. 208)

30ml (1fl oz) lime juice

cracked ice

spanked mint, to garnish

Shake all the ingredients together over cracked ice. Pour into an Old-Fashioned glass, adding the hull of the lime. Garnish with spanked mint.

*How can you get round the fact that the rum's unavailable? Beachbum Berry uses a 50:50 mix of an aged *agricole* and Appleton Estate 12-Year-Old. Using the funkier Smith & Cross, Bum told me, would be "like bringing a howitzer to a knife fight".

GOLDEN GLOVE

1 tsp white sugar

15ml (½oz) fresh lime juice

60ml (2fl oz) gold Jamaican rum

1 tsp Cointreau

340g (12oz) crushed ice

orange peel, to garnish

Dissolve the sugar in the lime juice, then blend all the ingredients with the ice for 20 seconds. Pour, unstrained, into a cocktail glass. Twist the orange peel above the glass to spray it with the citrus oil and then use the peel as a garnish.

MAI TAI

Trader Vic Bergeron knew his rum, so when in 1944 he grabbed a bottle of 17-year-old Wray & Nephew and mixed up a drink to give to his guests, Ham and Carrie Gould, it was quite deliberate. No wonder Carrie said *"Maita'i!"* (the Tahitian word for "good") when she tasted it. Pause for a second, though. A 17-year-old Jamaican rum in a cocktail? If you tried that these days, the drink would be £750 ($1,235). We know that, because that's what Belfast's Merchant Trader charged when it used up the last bottle.

To go back to the first point. Vic might have been a showman who liked to slam knives into his (false) leg to startle people, but he was also a businessman, and businessmen don't throw money away on expensive ingredients if there is an alternative around. He chose Wray & Nephew 17-year-old because it worked, and in the drink he allowed it to speak fully.

Vic and Donn Beach were the Bette Davis and Joan Crawford of tiki. Given this, it was inevitable that Donn claimed the Mai Tai as his own, or at least that it was a copy of his QB Cooler, though it was made with different ingredients. (Beachbum Berry, who knows these things, says the two taste the same). Did Vic try to stymie Donn, or did he think back to his days propping up the bar at Floridita watching Constante? After all, the Cuban great had made a Mai Tai-esque variant on his Daiquiri #2 drink called the Golden Glove.

Whatever the inspiration, it was Vic's drink – so much so that he kept the recipe secret all the way through the cocktail's glory days of the 1960s, when every bar was trying (and failing) to approximate the original.

Vic's secret was forced out of him in 1970 when he sued Donn Beach for claiming on the bottle of his Mai Tai mix to have been the creator of the drink. The fact that there were premixes available show both the ubiquity of the drink – and how far it had fallen. Thankfully, quality Mai Tai's have now become part of the new tiki revival. *Maurururu!* ("Thank you!")

GET YOUR MOJITO WORKING

Use the right glass. It's a refreshing slinger of a drink, not a pint.

Go easy on the mint. A couple of sprigs, not a bush. Don't pound it into submission with a baseball bat; that makes it taste like ditch water. A light press is all that's required.

Squeeze the lime. If you batter an innocent lime to death, you'll get bitter ditch water. It's juice you want.

Use simple (sugar) syrup unless you want to spent time dissolving the white sugar in the lime juice.

Use a flavoursome white rum: Havana Club, Caña Brava, Flor de Caña, Santa Teresa Claro. If you want to use an aged rum, then think about how it will affect the balance, and adjust.

1 tsp 2:1 simple syrup (*see* p.208)

juice of 1 lime

2 sprigs of mint (Note: It's not easy to get your hands on *hierba buena*, but if using spearmint don't muddle the stems.)

60ml (2fl oz) soda water

30ml (1fl oz) white rum

ice cubes

Angostura Bitters, optional

In a 240ml (8fl oz) glass, dissolve the simple syrup in the lime juice. Gently press leaves from one sprig of mint. Add the water. Stir. Add the rum, then the ice, give it all a gentle stir, then spank the second sprig of mint and use as garnish. Finally, go Cuban with a flick of Angostura Bitters, if desired.

MOJITO

By 1942, Angel Martinez was tired of life on the family farm in Villa Clara. He wanted to make his fortune in Havana and saw the modest wine shop cum grocery store (a *bodega*) in Calle Empredado as the solution. In the same year, he got a new neighbour in the print shop next door. Felito Ayon was a publisher who moved within the bohemian avant-garde milieu of Cuba, and when Martinez started to sell Créole food, Ayon's crowd began eating and drinking there. More people began to come to the little bodega in the middle of the block (El Bodeguita del Medio), which soon turned into Havana's equivalent of Paris' Les Deux Magots, but a hell of a lot more fun I'll wager.

A bar's reputation isn't just made by the quality of its drinks or service. It is made by the ambience created by its customers. In the 1940s, Cuba was full of great bars, all heaving with tourists and celebrities. The Bodeguita was everything they weren't. It was small, modest, and local. It was radical and subversive. It was a place where *bohemios* (wandering musicians) like Sindo Garay could sing their folk songs, where left-wing politics and art fused over sips of aged rum and a simple drink called the Mojito.

They were continuing a tradition of the mix of lime, mint, sugar, and spirit which allegedly soothed Sir Francis Drake (*see* p.13). A drink called the Draque was consumed across Spanish colonies until the start of the nineteenth century, though I wonder why a curative beverage would be called after a feared figure. It's like renaming Pepto-Bismol the Bogeyman.

More recently it became the Mai Tai of the Noughties, the default mixed drink for millennials. Don't get me wrong. I'm happy sipping a well-made Mojito and talking about avant-garde matters, but the former is less common than the latter. This is often a badly made drink. Send the bartenders for training at the Bodeguita del Medio? Sadly not. Go and eat there, drink aged rum, chat about intellectual matters, but steer clear of the Mojitos.

THE RECIPE

30ml (1fl oz) white rum (Havana
Club 3 Años, Caña Brava 3 Años)

....

30ml (1fl oz) gold rum

....

60ml (2fl oz) pineapple juice/
4 pineapple cubes

....

30ml (1fl oz) coconut cream

....

scant pinch of salt

....

crushed ice

....

Blitz all the ingredients together
with a cup of crushed ice until
smooth. Serve in a Collins glass.

If you don't have a blender, then
muddle the pineapple and shake.

Pineapple and rum just works.
"Pineapple rum" was more highly
priced at the end of the eighteenth
century than 10-year-old Jamaican.
The question is, was it flavoured,
or did the "pineapple" refer to an
estery note?

If aged rum is more your bag,
then proceed as follows:

PAINKILLER

120ml (4fl oz) Pusser's Gunpowder
Proof Rum

....

120ml (4fl oz) pineapple juice

....

60ml (2fl oz) coconut cream

....

60ml (2fl oz) orange juice

....

225g (8oz) crushed ice

....

Put all the ingredients in a blender
and blitz for three seconds. Serve in
the tiki mug closest to hand.

PIÑA COLADA

Piña Colada? I know what you're thinking. The ultimate
bad drink of the cocktail Dark Ages, that sickly sweet,
one-dimensional mess, consumed by the bucketload in
bars called things like "Slipper's" and "Romeo & Juliets"
[sic]. A disco drink, a "party" drink that deserved to be
sneered at by the sort of sophisticate who buys a book
like this.

I was the same until I stopped at the service station
next to Cuba's Puente de Bacunayagua. There was a bar.
Of course there was a bar; don't we all need a drink after
we've been driving for a few hours? It was also a bar
that only made one drink: a Piña Colada, using Havana
Club Blanco, fresh coconut, fresh pineapple, and sugar,
served in a hollowed-out coconut. It was awesome.
Prejudice fell away. My mind was opened.

It was appropriate that this Damascene conversion
took place in Cuba, as it was here that the drink had its
beginnings as a *piña fria*: fresh pineapple juice poured
over ice. It was one of the iced fruit drinks that became
popular at the start of the nineteenth century in taverns
such as the Piña de Plata. Add some *aguardiente* and
things became even better. Strain the juice and you
got a Piña Colada. By 1922, you could buy a mix of
pineapple juice, rum, ice, sugar, and lime juice in the bar
that had been born from the Piña de Plata, Floridita.

Coconut began to be added in the 1950s; a recipe for
a Cuban Piña Colada appeared in *The New York Times*
in 1950. But the drink as we know it first appeared in
Puerto Rico's Hilton Caribe hotel in 1954, immediately
after the first tinned coconut cream, Coco Lopez (also a
Puerto Rican product) was launched.

It's not a complex drink, but some depth can be
achieved by using gold or a mix of gold and white rums,
and leaving the double cream well alone. A tiny pinch of
salt, or a squeeze of lime also helps.

It remains a guilty pleasure, but only a single one can
be consumed. Any more and the sugar and the guilt
magnify, while the pleasure recedes.

EL PRESIDENTE MACHADO

30ml (1fl oz) Havana Club
Selección de Maestros

30ml (1fl oz) Dolin Blanc vermouth

2 tsp curaçao

1 tsp grenadine

Stir the ingredients together and
strain into a cocktail glass.

COMANDANTE CIENFUEGOS

40ml (1½fl oz) Havana Seleccion
de Maestros

20ml (¾fl oz) Martini Rosso

2 tsp fino Sherry

1 tsp green fig liqueur

lemon twist, to garnish

Stir the ingredients together and
strain into a coupette. Garnish with
a small lemon twist.

From Tony Conigliaro of Colebrook Row,
London.

EL PRESIDENTE

Eddie Woelke was a New York bartender who worked
behind the stick at the Biltmore hotel from 1913. When
John M Bowman of the Biltmore chain bought Havana's
Sevilla hotel (renaming it the Sevilla-Biltmore) in 1919,
Eddie headed south. It was here that he is said to
have invented the El Presidente in honour of Cuba's
president, Mario García Menocal.

In 1924, Eddie headed along to the American Jockey
Club and later to the Gran Nacional Casino in Marianao
(not the Hotel Nacional), creating a heap of fine drinks as
he went, including the Mary Pickford (see p.210). He also
tweaked El Presidente, adding curaçao, to salute Gerardo
Machado, the new occupant of that office in 1925.

I have an issue with this drink. Not the libation itself,
I hasten to add, which hits a beautiful balance between
the sweet and dry elements of aged Cuban rum, given
depth and length by the gentle, herbal, vinous quality
of blanc (not dry) vermouth. The curaçao adds lift and
links with the rum. It's a sophisticate – everything that
Machado wasn't, which is where my problem lies.

I don't hold Eddie to blame for this. If Machado asked
for a drink to be named after him, you'd probably jump
to it. This is, after all, a man under whose Prohibition-
era regime (1925–33) journalists were murdered and
opponents were fed to sharks. Graft and embezzlement
were endemic. He closed universities and high schools,
and courted the Mob. Do you really want to sip
something named after someone like that?

I reckon it's time for an improved El Presidente
named after Camilo Cienfuegos, the charismatic
revolutionary who once said, "Under no condition
should we put ourselves at the same moral level as
those we are fighting." He died in 1959 and is now,
sadly, forgotten. Camilo liked a drink, and deserves
his own as a small tribute. When I mentioned this to
Tony Conigliaro, another of his admirers, he came up
with the recipe left. No more need we be haunted
by Machado.

THE RECIPE

45ml (1½fl oz) gold rum

45ml (1½fl oz) aged Jamaican rum

30ml (1fl oz) Lemon Hart 151 Demerara

20ml (¾fl oz) fresh lime juice

15ml (½fl oz) Donn's mix * (see below)

15ml (½fl oz) Falernum

6 drops Pernod

1 tsp grenadine

dash Angostura Bitters

¾ cup crushed ice

cubed ice

mint sprig, to garnish

Dump it all in a blender and whizz for 5 seconds. Pour into a tall glass, adding ice cubes to fill. Garnish with a sprig of mint. Remember, two max!

* This is Donn's "Spices 4". Make a cinnamon syrup from three cinnamon sticks and one cup each of white sugar and water. Heat until the sugar is dissolved. Gently simmer for two minutes. Mix one part of this with two parts of white grapefruit juice.

Donn Beach, 1934

THE ZOMBIE

The wandering revenant has long troubled and thrilled the human mind. Although it may initially surprising that, in 1832, poet Robert Southey was the first to write the name "zombi", given that his friends included Mary Shelley, it's a good example of the dark Romantic state of mind. It wasn't until American troops occupied Haiti between 1915 and 1934 that various lurid tales, often masquerading as anthropology, were carried back. These included William Seabrook's zombie-fixated, voodoo "exposé", *Magic Island*. Seabrook, who claimed to have eaten human flesh, inspired a 1941 "voodoo" ceremony whose participants tried to kill Hitler by drinking Jamaican rum, banging drums, and ramming spikes into an effigy of the Führer.

The first zombie movie, *White Zombie*, appeared in 1932, while the considerably better *I Walked with a Zombie*, a Haiti-set, anti-slavery horror movie based on *Jane Eyre*, appeared in 1943 and is closer to the Haitian concept of a *zombi*: an animated corpse controlled by a *bokor* (wizard) to do his bidding. The parallels with slavery are obvious. In other words, in 1934, when Donn Beach stirred up a drink for a customer who promptly ordered another and then went out on a drunken spree, returning to Donn the day after, feeling like "the living dead", it was obvious what the new drink should be called.

Make no mistake: this is a strong drink. Two per customer, said Donn. After all, who wants a customer who is sliding down the wall after three glasses? The problem with the Zombie was the same that befell the Mai Tai: secrecy. Donn never let on what was in the original; all people knew was that it contained a lot of booze.

Its reputation as an annihilator is unfair to Donn Beach, who was an artist of mixed drinks. He didn't pick bottles at random, but carefully blended his ingredients to create complexity. We'd all still be guessing quite how the original was made, were it not for Beachbum Berry decoding a recipe in a notebook that belonged to Donn's former head bartender, Dick Santiago.

AIR MAIL (right)

60ml (2fl oz) gold rum (Barbados/St Lucia)

15ml (½fl oz) fresh lime juice

1 tsp honey

150ml (5fl oz) brut Champagne

cracked ice

Shake all the ingredients together over cracked ice, then pour, unstrained, into a Collins glass. Top up with Champagne.

From the 1949 *Handbook for Hosts*, and nominated by Naren Young.

CLASSIC COCKTAILS

CHET BAKER

Tim Philips says this modern classic, invented in 2005 by Sam Ross, is the rum drink he'd drink in "a nice joint", though as I've never seen that Aussie barfly in one, or been to one that would allow him in, the point is somewhat moot. The drink, however, shows his impeccable taste.

60ml (2fl oz) Rhum Barbancourt 5 Star Réserve Spéciale

1 tsp Punt e Mes

1 tsp honey syrup

2 dashes Angostura bitters

cubed ice

Shake all the ingredients together and strain into an ice-filled rocks glass.

FOGCUTTER

30ml (1fl oz) white rum

15ml (½fl oz) gin

15ml (½fl oz) VSOP brandy

60ml (2fl oz) fresh orange juice

30ml (1fl oz) fresh lemon juice

15ml (½fl oz) orgeat syrup

15ml (½fl oz) Amontillado sherry

1 mint sprig

cracked ice

Blend all the ingredients together with ice and serve in a suitably tiki-esque bowl.

Invented by Trader Vic Bergeron in the 1940s, and nominated by Tim Philips.

MARAGATO

30ml (1fl oz) Havana Club 3 años

15ml (½fl oz) dry vermouth

15ml (½fl oz) sweet vermouth

15ml (½fl oz) orange juice

15ml (½fl oz) lime juice

dashes of maraschino liqueur, to taste

cracked ice

Shake all the ingredients together with ice and strain into a chilled cocktail glass.

From *The Savoy Cocktail Book*, 1930.

2:1 SIMPLE SYRUP

Gently heat 2 parts white sugar to 1 part water until the sugar has completely dissolved. You can flavour the syrup by adding mint leaves, citrus peel, and so on. Alternatively, just buy a bottle of gomme.

HOTEL NACIONAL SPECIAL (right)

Created by Wil Taylor at Havana's Nacional Hotel (rather than by Eddie Woelke, who never worked there). Keep the rum light in nature and use fresh pineapple. A delicate treat.

45ml (1½fl oz) silver rum (or Havana Club 3 años)

45ml (1½fl oz) fresh pineapple juice

15ml (½fl oz) fresh lime juice

15ml (½fl oz) apricot brandy

cracked ice

lime or pineapple wedge, to garnish (optional)

Shake all the ingredients together with ice until very cold. Strain into a small cocktail stem glass. Garnish with lime or pineapple, if liked.

JUNGLE BIRD

45ml (1½fl oz) aged Jamaican rum

20ml (¾fl oz) Campari

20ml (¾fl oz) lime juice

20ml (¾fl oz) 2:1 simple syrup (*see* left) or gomme

45ml (1½fl oz) pineapple juice

ice cube

pineapple wedge, to garnish

Shake all the ingredients together and strain into a rocks glass with one ice cube. Garnish with a pineapple wedge.

First appeared in 1978 in the Aviary Bar in Kuala Lumpur Hilton. Nominated by Stuart McCluskey.

NAVY GROG

30ml (1fl oz) Demerara rum

30ml (1fl oz) aged Jamaican rum

**30ml (1fl oz) white Cuban/
Puerto Rican rum**

30ml (1fl oz) honey mix (*)

20ml (¾fl oz) fresh lime juice

20ml (¾fl oz) white grapefruit juice

20ml (¾fl oz) soda water

Shake the first six ingredients together. Strain into a glass and top with the soda.

The 1941 Donn Beach version, as nominated by Beachbum Berry.

*FOR DONN'S HONEY MIX

2 parts clover honey

1 part hot water

Blend until the honey has dissolved. Bottle and refrigerate.

MARY PICKFORD (right)

We were sitting in Floridita with London bartending legend Dick Bradsell. The bartender's hands are already poised to make a Daiquiri. Dick says, "One moment, please. I'd like a Mary Pickford." The bartender steps back, smiles, and crafts a classic drink. Thanks, Dick.

45ml (1½fl oz) Havana Club 3 Años

30ml (1fl oz) fresh pineapple juice

2 dashes maraschino liqueur

1 dash grenadine

crushed ice

1 maraschino cherry, to garnish

Either blend all the ingredients together with crushed ice and serve unstrained, or shake them together and strain. Garnish with a maraschino cherry.

MULATA DAIQUIRI

A grown-up Daiquiri and one of my favourite rum drinks.

45ml (1½fl oz) Havana Club Añejo 7 Años

30ml (1fl oz) Tempus Fugit crème de cacao

15ml (½fl oz) fresh lime juice

1 tsp sugar

crushed ice

Either blend all the ingredients together with crushed ice and serve unstrained, or shake them together and strain.

CARTA SWITCHEL (right)

20ml (¾fl oz) Bacardí Carta Blanca

20ml (¾fl oz) Bacardí Bacardí Oro

20ml (¾fl oz) grapefruit juice

10ml (⅓fl oz) gomme syrup

4 dashes apple vinegar

4 dashes Fernet-Branca

crushed ice

grapefruit zest, to garnish

mint sprig, to garnish

Churn all the ingredients together over crushed ice in a highball glass. Garnish with grapefruit zest and a mint sprig.

From Iain Griffiths of Dandelyan, London.

CLOUDS ABOVE THE CANE

40ml (1½fl oz) Mount Gay Black Barrel

30ml (1fl oz) Pineau des Charentes blanc

1 tsp gin

candyfloss, to garnish

Stir all the ingredients together and serve in a tasting glass with a candyfloss garnish.

From Stuart McCluskey of Bon Vivant, Edinburgh.

MODERN TWISTS

AUTUMNAL PUNCH

30ml (1fl oz) Scarlet Ibis rum (if unavailable, blend Angostura 5-year-old rum and heavy pot-still or aged *agricole*)

30ml (1fl oz) Marie Brizard Poire William

30ml (1fl oz) lemon juice

1 tsp St Elizabeth Allspice Dram

2 tsp orgeat syrup

cracked ice

raisins (preferably golden)

Combine all the ingredients except for the raisins in a mixing glass and fill with ice. Fill a tall glass with fresh ice and sprinkle the raisins through. Shake for 10 seconds, then strain over the fresh ice.

From H. Joseph Ehrmann of Elixir, San Francisco.

COCOA REPUBLIC

40ml (1½fl oz) white rum

20ml (¾fl oz) Noilly Prat

1 tsp orange liqueur

1 tsp white crème de cacao

5ml (⅙fl oz) grenadine

cracked ice

orange twist, to garnish

Stir all the ingredients together over ice. Strain into a dainty glass and garnish with an orange twist to serve.

From Robin Honhold of White Lyan, London.

CUBAN MIST

45ml (1½fl oz) Havana Club 7 Años

20ml (¾fl oz) apricot brandy

20ml (¾fl oz) Hennessy Cognac

1 tsp muscovado syrup

2 dashes Bitter Truth Old Time Aromatic Bitters

orange zest, to garnish

Stir all the ingredients together and strain into a coupette. Garnish with orange zest to serve.

From Tess Postumus of Amsterdam.

GREEN THUMB (right)

60ml (2fl oz) Caña Brava 3 Años rum

15ml (½fl oz) lime juice

7.5ml (¼fl oz) St Germain elderflower liqueur

7.5ml (¼fl oz) celery juice

7.5ml (¼fl oz) 2:1 simple syrup (see p.208)

⅛ tsp matcha powder

cracked ice

cucumber slice, to serve

Shake all the ingredients together first to incorporate the matcha powder, then shake again with ice and fine-strain into a chilled coupette. Garnish with a slice of cucumber to serve.

From Jim Meehan of mixographyinc.com.

EQUINOX

22.5ml (¾fl oz) white Virgin Islands rum

22.5ml (¾fl oz) gold Virgin Islands rum

15ml (½fl oz) fresh lime juice

15ml (½fl oz) Falernum

15ml (½fl oz) honey syrup (1:1 ratio honey to water)

22.5ml (¾fl oz) coconut milk

ice cubes

long strip of channel-cut lime peel, twisted into a spiral, to garnish

Put all the ingredients into a cocktail shaker and shake together with ice cubes. Strain over fresh cubes into a rocks glass or speciality glass. Garnish with the lime peel.

From Jeff Beachbum Berry of Latitude 29, New Orleans.

HATS OFF TO BERRY

50ml (1¾fl oz) rum blend (Cuba, Jamaica, Barbados)

20ml (¾fl oz) berry syrup (such as blueberry or blackberry)

30 (1fl oz) lime juice

1 dash absinthe

2 dashes Angostura Bitters

lime wheel, mint sprig, passion fruit slice or orchid, to garnish

Pour everything into a shaker. Shake hard and strain into a rocks glass. Garnish with a lime wheel, mint sprig, passion fruit slice, or orchid.

From Scotty Schuder of Dirty Dick, Paris.

JOSÉ MARTI SPECIAL

4 cloves

40ml (1⅓fl oz) Havana Club 3 Años

½ tsp Ricard

15ml (½fl oz) Tio Pepe Sherry

20ml (¾fl oz) lime juice

20ml (¾fl oz) 2:1 simple syrup (*see* p.208)

cracked ice

Muddle the cloves in the bottom of the shaker. Add the remaining ingredients, shake together with ice, and strain in a coupette. Serve with a straw.

From Andy Loudon, winner of the Havana Club Grand Prix 2014.

KAIETEUR SWIZZLE

60ml (2fl oz) El Dorado 8-Year-Old or 12-Year-Old

22ml (¾fl oz) fresh lime juice

15ml (½fl oz) grade A maple syrup

15ml (½fl oz) John D Taylor's Velvet Falernum Liqueur

2 dashes Angostura Bitters

crushed ice

mint sprig, to garnish

Add all the ingredients to a Collins or highball glass and fill the glass three-quarters full with crushed ice. Swizzle with a swizzle stick or bar spoon. Top up with additional crushed ice as needed to fill the glass, and garnish with a mint sprig.

From Martin Cate of Smuggler's Cove, San Francisco.

LATIN QUARTER

crushed ice and cracked ice

60ml (2fl oz) Ron Zacapa Centenario Sistema Solera 23

1 tsp rich sugar cane syrup

3 dashes Peychaud's Bitters

1 dash Angostura Bitters

1 dash Bittermens Xocolatl Mole Bitters

Spray of Pernod Absinthe

lemon twist

Chill an Old-Fashioned glass with crushed ice. Stir all the ingredients except the absinthe together over cracked ice. Remove the crushed ice from the Old-Fashioned glass and rinse with 2–3 dashes of the absinthe, turning to coat. Strain the cocktail into a chilled, rinsed glass and squeeze a lemon twist over the drink. Discard the peel and serve.

From Joaquín Simó of Pouring Ribbons, New York, New York.

MAID IN CUBA

absinthe rinse

cubed ice

50ml (1⅔fl oz) Bacardí Carta Blanca

30 (1fl oz) lime juice

20ml (¾fl oz) 2:1 simple syrup (*see* p.208)

3 slices of cucumber, plus extra to garnish

6 mint leaves

15ml (½fl oz) soda

Pour a dash of absinthe into an ice-filled coupette and top with water. Shake all the other ingredients except the soda together. Discard the contents of the cocktail glass and strain the contents of the shaker into it. Spritz with soda water and garnish with a slice of cucumber.

From Tom Walker, UK, global winner 2014 Bacardí Legacy competition.

RAY BARIENTOS

50ml (1⅔fl oz) aged medium
bodied rum (Bacardí 8-Year-Old,
Zacapa 15 Solera, Angostura 1919)

30ml (1fl oz) fresh lime juice

15ml (½fl oz) fresh orange juice

2 tsp cherry brandy

2 tsp cinnamon syrup

2 dashes Angostura Aromatic Bitters

cinnamon stick, to garnish (optional)

Shake all the ingredients together
and double-strain into a coupette
glass. Garnish with a cinnamon
stick if desired.

From Thanos Prunarus of Baba Au
Rum, Athens.

LIN BABA DAIQUIRI

45ml (1½fl oz) The Scarlet Ibis Trinidad Rum

15ml (½fl oz) Appleton Estate Signature Blend

7.5ml (¼fl oz) orgeat syrup

7.5ml (¼fl oz) ginger syrup

20ml (¾fl oz) lime juice

1 dash Angostura Bitters

2 green curry leaves (do not muddle)

7g (¼oz) cinnamon bark

Shake all the ingredients together and fine-strain into
a cocktail glass.

From Joaquín Simó of Pouring Ribbons, New York, New York.

LE LATIN

45ml (1½fl oz) Bacardí
Carta Blanca

20ml (¾fl oz) Viognier
white wine

20ml (¾fl oz) lemon juice

2 tsp olive brine

2 tsp caster sugar

1 olive, to garnish

Shake all the ingredients together
and strain into a cocktail glass.
Garnish with an olive.

From Franck Dideu, France, global winner
of the 2015 Bacardí Legacy competition.

YUZU BREEZER

30ml (1fl oz) Bacardí Carta Blanca

20ml (¾fl oz) Yuzushu

2 tsp Aperol

dash plum bitters

small pinch citric acid

70ml (2½fl oz) soda water

Shake the first five ingredients together and strain.
Add the soda water. It's served in a bottle at the bar,
but you could use a glass.

From Tim Philips of Bulletin Place, Sydney.

COCKTAILS

THE LOST LOVERS

60ml (2fl oz) aged Venezuelan, or other medium-bodied aged, rum

2 tsp Pedro Ximénez Sherry

2 dashes orange flower water

2 dashes teapot bitters

the extracted oil from the peel of 2 oranges (*see* "oleo saccharum" under Regent's Punch, p.191)

cubed ice

Stir all the ingredients in a double Old-Fashioned glass with a couple of nice dry, big ice cubes.

From Thanos Prunarus of Baba Au Rum, Athens.

OLD GOLD (right)

30ml (1fl oz) Ron Zacapa

20ml (¾oz) Caol Ila whisky

1 tsp agave nectar

1 tsp Tempus Fugit Gran Classico Bitters

pinch sea salt

cacao nibs, to garnish (optional)

Shake all the ingredients together and strain into a rocks glass. Grate cacao nibs over the top if you want a garnish.

From Tim Philips of Bulletin Place, Sydney.

MOLASSACRE PUNCH

Named for the great Boston molasses flood of 1919. A massive tank at a distillery burst open, spilling out millions of gallons of molasses and creating a 8-metre (25-foot) wave that reached speeds of 48kph (30mph). Dozens were taken out in the wave and hundreds hurt. Some say that, on a hot day, the sweet smell still lingers. Take it as you will, but it happened a day before Prohibition was voted in.

45ml (1½fl oz) Smith & Cross Traditional Jamaica Rum

20ml (¾fl oz) Rioja

15ml (½fl oz) green tea and ginger syrup (mix 2 tsp freshly brewed green tea with 1 tsp ginger syrup)

20ml (¾fl oz) lemon juice

1 tsp currant jelly

cracked ice

lemon twist, to garnish

Combine all the ingredients in a tin or cocktail shaker and shake well with ice. Fine-strain into a rocks glass. Garnish with a lemon twist to serve.

From Charles Joly of crafthousecocktails.com.

L'ORANGE DROP DAIQUIRI

50ml (1⅔fl oz) Saint James Rhum Agricole Blanc

30ml (1fl oz) fresh lime juice

12.5ml (⅓fl oz) Agave Rèal Blue Agave Nectar

3 dashes Angostura Orange Bitters

oils from a twist of orange peel, to garnish

Shake all the ingredients together and serve in a coupette. Garnish with the oils from a twist of orange peel (*see* "oleo saccharum" under Regent's Punch, p.191).

From Ian Burrell of Cotton's, London.

COCKTAILS

ST JAMES GATE

50ml (1⅔fl oz) Myers's

30ml (1fl oz) lemon Juice

30ml (1fl oz) egg white

**15ml (½fl oz) 2:1 simple syrup
(*see* p.208)**

**20ml(¾fl oz) Guinness reduction:
50g (1⅙oz) treacle to 500ml
(18fl oz) of Guinness reduced
on a low heat until it is half its
original volume**

cracked ice

Dry shake all the ingredients
together, then shake over ice
and serve in a large coupette.

From Tony Conigliaro of 69 Colebrook
Row, London.

RUM & RAISIN FLIP

45ml (1½fl oz) Appleton Estate Reserve Jamaica Rum

15ml (½fl oz) Pedro Ximénez Sherry

15ml (½fl oz) Frangelico

3 dashes Dale DeGroff's Pimento Aromatic Bitters

1 whole egg

cracked ice

grated nutmeg, to serve

Shake all the ingredients together very hard with ice.
Strain into a goblet. Grate nutmeg on top to serve.

From Naren Young of Dante, New York, New York.

COCKTAILS

MAIN SOURCE BIBLIOGRAPHY

Abbott, Elizabeth. *Sugar: A Bittersweet History*. London: Duckworth Publishers, 2009.

Allchin, F R. "India: The Ancient Home of Distillation?", in *Man* (Vol. 14, No. 1, Mar. 1979), pp. 55–63. London: Royal Anthropological Institute of Great Britain and Ireland, 1979.

Asbury, Herbert. *The Great Illusion: An Informal History of Prohibition*. Garden City, New York: Doubleday, 1950.

Ayala, César J. *American Sugar Kingdom: The Plantation Economy of the Spanish Caribbean, 1898–1934*. Chapel Hill, North Carolina: University of North Carolina Press, 1999.

Barty-King, Hugh, and Massel, Anton. *Rum: Yesterday and Today*. London: William Heinemann Ltd, 1983.

Belgrove, William. *A Treatise Upon Husbandry or Planting*. Boston: D. Fowle, 1755.

Berry, Jeff. *Beachbum Berry's Potions of the Caribbean*. New York: Cocktail Kingdom, 2014.

British Guiana Administration Reports. Georgetown, Demerara: The Argosy Co., 1905.

Brown, Jared, and Miller, Anistatia. *Cuban Cocktails*. London: Mixellany Ltd, 2012.

Brown, Jared, and Miller, Anistatia. *Spiritous Journey: A History of Drink, Books 1 and 2*, London: Mixellany Ltd, 2009.

Bolingbroke, Henry. *A Voyage to the Demerary*. London: Richard Phillips, 1807.

Bonera, Miguel. *Oro Blanco Tomo 1*. Toronto: Lugus Libros, 2000.

Bose, Dhirendra Krishna. *Wine in Ancient India*. Calcutta: K. M. Connor & Co, 1922.

Brown, John Hull. *Early American Beverages*. Rutland, Vermont: C E Tuttle Company, 1966.

Bruno, Sergio Nicolau Freire. "Distillation of Brazilian Sugar Cane Spirits (Cachaças)" in *Distillation: Advances from Modeling to Applications*, Dr Sina Zereshki (Ed.), ISBN: 978-953-51-0428-5, InTech. Available from: www.intechopen.com.

Camard-Hayot, Florette and Laguarigue, Jean-Luc de. *Martinique Terre de Rhum*. Bordeaux: Traces H.S.E., 1997.

Campoamor, Fernando G. *Hemingway's Floridita*. Toulouse: Editions Bahia Presse.

Cooper, Ambrose. *The Complete Distiller*. London, 1757.

Coulombe, Charles A. *Rum: The Epic Story of the Drink That Conquered the World*. New York: Citadel Press, 2004.

Curtis, Wayne. *And a Bottle of Rum: A History of the New World in Ten Cocktails*. New York: Three Rivers Press, 2007.

Daniels C, Needham J, and Menzies, Nicholas K (eds). *Science and Civilisation in China*, Volume 6. Cambridge, UK: Cambridge University Press, 1996.

Eaden, J. *The Memoirs of Père Labat 1693–1705*. London: Frank Cass & Co. Ltd., 1970.

Edwards, Bryan. *The History, Civil and Commercial, of the British West Indies: Vol. 2*. London: John Stockdale, 1819.

Fawcett, William. *Bulletin of the Botanical Department, Vol III*. Jamaica: Kingston Botanical Department: 1896.

Forbes, R J. *A Short History of the Art of Distillation from the Beginnings up to the Death of Cellier Blumenthal*. Leiden, Netherlands: E. J. Brill, 1970.

Foss, Richard. *Rum: A Global History*. London: Reaktion Books Ltd., 2012.

García Pepín, Anabel. *Rum in Puerto Rico: Tradition and Culture*. San Juan: Rones de Puerto Rico, Compañía de Fomento Industrial, 2002.

Haigh, Ted. *Vintage Spirits and Forgotten Cocktails*. Beverly, MA: Quarry Books, 2009.

Hearn, Lafcadio. *Two Years in the French West Indies*. Teddington, Middlesex, UK: Echo Library. 2006.

Hoarau, Michel. *Rhum (Le) de île de La Réunion*. Réunion: private press, 2001.

Huetz de Lemps, A. *Histoire du Rhum*. Paris: Editions Desjonquères, 1997.

Hui, Y H, Evranuz, E (eds). *Handbook of Plant-Based Fermented Food and Beverage Technology*. Oxford, UK: CRC Press Abingdon, date unknown.

Kieschnick, John. *The Impact of Buddhism on Chinese Material Culture*. Princeton, New Jersey: Princeton University Press, 2003.

Knight, Franklin W. *The Caribbean*. New York: Oxford University Press, 1990.

Kobler, John. *Ardent Spirits: The Rise and Fall of Prohibition*. New York: Da Capo Press Inc., 1993.

Lam, Rafael, and Bowler, Tim. *The Bodeguita del Medio*. Havana: Editorial José Marti, 1999.

Ligon, Richard. *A True and Exact History of the Island of Barbados*. Bath, UK: Bookcraft, 1998.

Martin, Samuel. *An Essay upon Plantership*. Antigua: Samuel Jones, 1756.

Mintz, Sidney W. *Sweetness and Power: The Place of Sugar in Modern History*. London: Penguin Books, 1986.

Morewood, Samuel. *Philosophical and Statistical History of the Inventions and Customs of Ancient and Modern Nations in the Manufacture and Use of Inebriating Liquors*. Dublin: William Curry, Jnr, and Company, 1838.

Niazi, Ghulam Sarwar Khan, Dr. *The Life and Works of Sultan Alauddin Khalji*. New Delhi, India: Atlantic Publishers, 1992.

O'Connell, Sanjida. *Sugar: The Grass that Changed the World*. London: Virgin Books, 2004.

Ortiz, Fernando. *Cuban Counterpoint: Tobacco and Sugar*. Durham, North Carolina: Duke University Press, 1995.

Parker, Matthew. *The Sugar Barons: Family, Corruption, Empire and War*. London: Random House, 2011.

Pack, James, Captain. *Nelson's Blood: The Story of Naval Rum*. Stroud, Gloucestershire, UK: Allan Sutton Publishing Ltd., 1995.

Pérez Jr, Louis A. *On Becoming Cuban: Identity, Nationality, and Culture*. Chapel Hill, North Carolina: University of North Carolina Press, 1999.

Report on the Experimental Work. Kingston, Jamaica: Sugar Experiment Station, 1906.

Roberts, Justin. *Slavery and the Enlightenment in the British Atlantic, 1750–1807*. Cambridge, UK: Cambridge University Press, 2013.

Shamasastry, R. *Kautilya's Arthashastra Translated into English*. Bangalore: 1915.

Sheridan, Richard B. *Sugar and Slavery: An Economic History of the British West Indies, 1623–1775*. Barbados, Jamaica: University of the West Indies, Canoe Press, 2007.

Smith, Frederick H. *Caribbean Rum: A Social and Economic History*. Gainesville, Florida: University Press of Florida, 2008.

Sublette, Ned. *Cuba and Its Music: From the First Drums to the Mambo*. Chicago: Chicago Review Press, 2004.

Taussig, Charles William. *Rum Romance and Rebellion*. London: Jarrolds, date unknown.

Thompson, Peter. *Rum Punch and Revolution*. Philadelphia: University of Pennsylvania Press, 1999.

Verhoog, Jeroen. *Walking on Gold*. Amsterdam: E&A Sheer, 2013.

Weeden, William. *Economic and Social History of New England, 1620–1789*. Cambridge, Massachusetts: Houghton, Mifflin, 1890.

Williams, Ian. *Rum: a Social and Sociable History*. New York: Nation Books, 2005.

Wondrich, David, *Punch: The Delights (and Dangers) of the Flowing Bowl*. New York: Penguin Group, 2010.

Wray, Leonard. *The Practical Sugar Planter*. London: Smith, Elder & Co., 1848.

Y-Worth, W. *The Compleat Distiller*. London: J Taylor, 1705.

INDEX

THANKS

PICTURE CREDITS

The publishers would like to thank all the rum makers, distributors and agents who have kindly provided images for use in this book.

Special photography for Octopus Publishing: **Cristian Barnett**

Additional credits are as follows.

Alamy Stock Photo Didier Forray/Sagaphoto.com 33; age fotostock 48; Everett Collection 13; Falkenstein/Bildagentur-online Historical Collect 24; Florilegius 11; GL Archive 12; Guy Harrop 55; Pulsar Images 51. Courtesy **Angostura** 46.

Bridgeman Images Pictures from History 31; The Stapleton Collection 21.

© **Decca** Records, 1945 34.

Courtesy **The Duppy Share** 2, 8, 9, 38, 44.

Getty Images Adalberto Roque/AFP 41; Franck Guiziou 39; Jim Heimann Collection 35; Jonathan Blair 42; MPI 18; Nelson Almeida/AFP 52; Spencer Arnold 27; Steve Russell/Toronto Star 57; Universal History Archive/UIG 19.

Courtesy **Haus Alpenz** 53.

istockphoto.com nengredeye 40.

Courtesy Gayle Seale/**R L Seal** 45.

Scotchwhisky.com 7.

SuperStock Hemis 49.

TopFoto 14; EUFD 16, 26; The Granger Collection 23; The Print Collector/HIP 30; World History Archive 10.

Via **vintageadbrowser.com** 29.

© **The Whisky Exchange** 1999-2016 All Rights Reserved 71, 74, 140, 149, 155, 156, 164, 170, 184, 186.

To Carsten Vlierboom, who has calmly taught me more about rum blending than anyone. Richard Seale, for his endless patience, opinions, and technical input. Luca Gargano, with whom it is always a treat to share a stage. Ian Burrell for his boundless passion in the cause of rum – I only went to bed for one hour! Jeff "Beachbum" Berry and Annene, for all the tiki info, Asbel Morales for opening up the world of Cuban rum, and François Renie for doing the same with Cuban music. Bruce Perry and John Barrett: old rum hands always supportive; also to Ryan Cheti, Laurent Broc, Lorena Vasquez, Rebecca Quiñonez, Christelle Harris, Alexandre Gabriel, Meimi Sanchez, Nick Blacknell, Chris Middleton, Alejandro Bolivar, Shervene Shahbazkhani, the Burrs; Keir, Arthur, and The Deacon [RIP] for allowing me to preach in Da Rum Chapel; Stef, who still has molasses in her heart; Steve Hoyles and Tony Hart for showing me the way.

To all those who helped with samples and info: Gabrielle D'Alessandro, Lauren Bajdala Brown, Sonia Bastian, Florent Beuchet, Daniele Biondi, Carl Blackwell, Edward M. Butler, Agatha Chapman-Poole, Oliver Chilton, Ashok Chokalingam, Gabrielle Cole, Emma Currin, Claire Desnoyer, George Frost, Simon Ford, Tom Gamborg, Jenny Gardener, Jessica Gibbons, Nick Gillett, Charlie Graham, the great Jim Grierson, Chris Hysted, Marissa Johnston, Pavol Kazimir, Alexander Kong, Nathalie de Labrouhe, Nicolas Legendre, Duncan Littler, Catherine McDonald, Gregory Neisson, Su-Lin Ong, Bailey Pryor, Fabio Rossi, Chris Seale, Jess Swinfen, Luke Tegner, Cynthia Thomas, Peter Thornton, Guy Topping, Tarja Tuunanen, Abbigale Wallis, Dan Warner, Larry and Simon Warren, Sarah Watson, John West, Emily Wheldon, Lexi Winsley, and to Raj, Dawn, and Jacqui at TWE, for sourcing some obscure rums and prompt delivery. Keshav, for the McDowell's, and JD, for breaking with his jet-set lifestyle and buying some Tanduay in a Manila supermarket.

To all the bartenders: Martin Cate, Tony Conigliaro, H Joseph Ehrmann, Iain Griffiths, Robin Honhold, Charles Joly, Andy Loudon, Stu McCluskey, Jim Meehan, Tim Philips, Tess Postumus, Thanos Prunarus, Scotty Schuder, Joaquín Simó, Tom Walker, Naren Young, and thanks to all who, over the past decade, have taken the time to ask when the next rum book was coming out. Hope you like this one.

My fellow bloggers and scribes: Wayne Curtis, Simon Difford, Martine Nouet, Stashe and Jared, Chris Middleton, John Gibbons, Ian Williams, David Wondrich, and those who lurk behind the following great blogs: thelonecaner.com, rumdood.com, thefatrumpirate.com, thefloatingrumshack.com, www.cachagora.com, robsrumguide.com, www.bostonapothecary.com.

To Denise, Leanne, Juliette, Giulia, Jamie, Geoff, Liz, and all at Octopus: it has been like swimming through molasses at times, so thanks for your belief, guidance, remarkable patience, and, again, fantastic work and dedication; and to my agent Tom Williams for wise words of reason.

To my wonderful wife, Jo, who stayed preternaturally calm when dealing with myriad sample requests and seemingly tedious but actually incredibly important bits of micro-managing that go into making my life easier. And finally, to my daughter, Rosie, who can now add Ace Daiquiri Wrangler to her increasingly impressive credentials. I love you both.